UNDERSTANDING THE MYSTERIES OF PRACTICAL HUMAN BIOLOGY
for
NON-SCIENTISTS

FIRST EDITION

UNDERSTANDING THE MYSTERIES OF PRACTICAL HUMAN BIOLOGY
FOR
NON-SCIENTISTS

JEFFREY A. KUSHNER, Ph.D.

Bassim Hamadeh, CEO and Publisher
Alisa Munoz, Project Editor
Celeste Paed, Associate Production Editor
Jess Estrella, Senior Graphic Designer
Greg Isales, Licensing Associate
Natalie Piccotti, Director of Marketing
Kassie Graves, Vice President of Editorial
Jamie Giganti, Director of Academic Publishing

Copyright © 2021 by Cognella, Inc. All rights reserved. No part of this publication may be reprinted, reproduced, transmitted, or utilized in any form or by any electronic, mechanical, or other means, now known or hereafter invented, including photocopying, microfilming, and recording, or in any information retrieval system without the written permission of Cognella, Inc. For inquiries regarding permissions, translations, foreign rights, audio rights, and any other forms of reproduction, please contact the Cognella Licensing Department at rights@cognella.com.

Trademark Notice: Product or corporate names may be trademarks or registered trademarks and are used only for identification and explanation without intent to infringe.

Cover image copyright © 2014 Depositphotos/orlaimagen.

Printed in the United States of America.

3970 Sorrento Valley Blvd., Ste. 500, San Diego, CA 92121

CONTENTS

Introduction	1
	1

LESSON 1
Homeostasis—The Goal of Human Biology — 3

Objectives	3
Defining Homeostasis	3
An Example of Homeostasis	3
Individual Body Systems, Their Functions, and Homeostasis	4
Negative and Positive Feedback Systems and Homeostasis	5
Another Example of Negative Feedback	6
Key Points	6
Reference	7

LESSON 2
The Mysteries of Science—Scientific Method versus Serendipity — 8

Objectives	8
Science, the Scientific Method, and Serendipity	8
Key Points	9
Reference	9

LESSON 3
The Mysterious Bond between Chemistry and Human Biology — 10

Objectives	10
Atoms	11
Atoms Are the Basis for Everything in the Universe	12
Radioactive Atoms	13
Electrons	13
Bonding	14
Acids and Bases	15
The Major Organic Macromolecules	16
Carbohydrates	16
Proteins	16

Lipids	17
Nucleic Acids	19
Key Points	19

LESSON 4
The Mysteries of Structure — 23

Objectives	23
Cells	24
The Cell Theory	24
Cells and Cellular Organization	24
The Cell Membrane	25
Transport into and out of Cells	25
Water and Solute Movement	26
Cells and Diffusion	26
Tonicity and Osmosis Definitions	26
Passive, Carrier-Assisted, and Active Transport	27
Vesicle-Mediated Transport	27
Cell Organelles	28
The Cell Wall	28
Cytoplasm	28
Mitochondria	28
Enzymes	29
Factors That Influence Enzyme Activity	29
Metabolism: The Transformation of Energy	30
Anabolic Reaction (Anabolism)	30
Catabolic Reaction (Catabolism)	31
Carbohydrate Catabolism	31
Aerobic Cellular Respiration	32
Glycolysis	32
How Do We Metabolize Things That Aren't Glucose?	33
Metabolism: Diet and Nutrition	33
What Is a Calorie?	34
Carbohydrates	34
Proteins	35
Dietary Lipids	35
Micronutrients	36
The Role of the Cytoplasm and Mitochondria in Cellular Metabolism and Fermentation: Recap	36
Mitochondria	36
Glycolysis (Lysing Glucose)	37
Anaerobic Pathways—Fermentation	37
Aerobic Respiration	37
The Endomembrane System	37
Vacuoles and Vesicles	38
Lysosomes	38
Ribosomes	38
Endoplasmic Reticulum	38
Golgi Apparatus	38

Information Management—The Nucleus	39
Cell Reproduction	39
Mitosis	40
Meiosis	40
Phases of Meiosis	40
Cell Division—Mitosis and Meiosis: Summary	41
Mitosis	41
Meiosis	43
Sexual Reproduction	43
Chromosomes and Genes	44
Terminology	44
DNA and Molecular Genetics	44
Protein Synthesis	45
One-Gene-One-Protein (This is the old theory, but is it still true?)	45
Mutations Redefined	45
Control of Gene Expression	45
The Eukaryotic Genome	45
Genes, Viruses, and Cancer	46
Molecular Genetics—DNA Replication and Gene Expression: Summary	46
DNA Replication—Copying of a Double-Stranded DNA Molecule	46
Replication Mistakes: Mutations of Genes	47
Gene Expression: Transcription and Translation (Making Proteins)	47
Genetic Information Copied from DNA Is Transferred to Three Types of RNA	48
Transcription	48
Translation	48
Transcription and Translation: Summary	49
Example Hereditary Disease Due to DNA Mutation (Sickle Cell Anemia)	50
What Is Epigenetics?	50
Genetics (Heredity)	50
Mendel and His Peas	50
Genetic Terms	51
Sex-Linkage of Genes	51
The Modern View of the Gene	51
Genes and Chromosomes	51
Key Points	52

LESSON 5
Function of Skeleton and Joints   57

Objectives	57
Skeletal System	58
The Axial and Appendicular Skeletons	58
Axial Skeleton	59
Appendicular Skeleton	62
Bone Tissue	63
Bone Growth	63
Joints	63
Key Points	63

LESSON 6
The Mysteries of Human-FunctionSystems — 66

Objectives — 66
Musculoskeletal Systems — 66
Skeletal Muscle Structure — 67
Control of Muscle Contraction — 67
Key Points — 68

LESSON 7.1
Imagine the Possible Brain Functions — 70

Objectives — 70
Central Nervous System and the Peripheral Nervous System — 71
The Nervous System — 71
The Neural Message—Action Potential (An "All-or-None" Phenomenon) — 71
    Generation of an Action Potential Reviewed — 72
Synapses—Electrical to Chemical Transmission of the Action Potential Signal — 72
Nervous System Functions — 73
    Sensory Input — 73
    Integration and Output — 73
Divisions of the Nervous System — 73
    The Peripheral Nervous System (PNS) — 74
    The Somatic Nervous System (SNS) — 74
    The Autonomic Nervous System (ANS) — 74
    The Central Nervous System (CNS) — 74
Key Points — 75

LESSON 7.2
Special Senses — 79

Objectives — 79
Sensation and Sensory Receptors — 79
Special Senses — 79
    Vision (Photoreceptors and Light) — 81
    Hearing — 81
    Orientation and Gravity — 81
Key Points — 81

LESSON 8
Hormonally Speaking—The Endocrine System — 84

Objectives — 84
Hormones — 84
Endocrine Systems and Feedback Cycles — 85
Hormone Action and Potential Problems — 85
The Endocrine System — 86

The Hypothalamus and Pituitary Gland	86
The Other Endocrine Organs	87
The Thyroid Gland	87
The Parathyroid Glands	87
The Pancreas	87
The Adrenal Glands	88
Key Points	88

LESSON 9.1
Heart-y Functions and Blood — 91

Objectives	91
The Heart	91
The Circulatory System	93
The Vessels of the Vascular System	94
Diseases of the Heart and Cardiovascular System (Mostly Due to Hypertension or Atherosclerosis)	94
Key Points	95

LESSON 9.2
Blood — 97

Objectives	97
The Blood	97
Three Types of Formed Elements	98
Three Types of Granulocytes	98
Two Types of Agranulocytes	98
Key Points	99

LESSON 10
Inspirational Breathing Functions — 102

Objectives	102
The Respiratory System and Gas Exchange	102
Key Points	104

LESSON 11
Sustaining Life Functions—Digestion and Nutrition — 106

Objectives	106
Overview	107
Stages in the Digestive Process	108
Components of the Digestive System	108
The Mouth, Pharynx, and Esophagus—Some Mechanical and Chemical Digestion	108
The Stomach—More Mechanical and Minor Digestion	109
The Small Intestine	109
The Large Intestine	110

Regulation of Appetite	110
Nutrition	110
Nutrition Labels	111
Health Claims	112
Key Points	112

LESSON 12
Waste Management Functions — 115

Objectives	115
The Waste Management System	116
The Human Excretory System	117
Hormones and the Kidneys	117
Kidney Stones and Other Renal Problems	118
Key Points	118
Summary	120

LESSON 13
Male and Female Sexual Functions — 121

Objectives	121
Reproductive System and Gametogenesis	122
Endocrine Control	123
Anatomy of the Male Reproductive System	123
Anatomy of the Female Reproductive System	123
Egg Production Takes Place within the Ovaries	124
Key Points	124

LESSON 14
Protective Immune Functions — 127

Objectives	127
The Lymphatic and Immune Systems	127
Immunity	128
General Defenses	128
Specific Defenses	129
Macrophages	129
B Cells	129
T Cells	130
The Immune System and Memory	130
Blood Types, Rh, and Antibodies	130
Organ Transplants and Antibodies	131
Allergies and Disorders of the Immune System	131
Summary	132
Lymphatics/Lymph Nodes	132
Antigens	132
First Line of Defense—Nonspecific	132

Second Line of Defense—Nonspecific	132
Components of the Second Line of Defense	132
Third Line of Defense—Acquired/Specific	133
Antibodies	133
Key Points	134

LESSON 15
Concluding Remarks 137

Introduction

Book Objectives

The successful human biology student reading this book should be able to accomplish the following:

- **Demonstrate** an understanding of human biology.
- **Describe** the structure and function of the many systems.
- **Explain** how the body maintains coordination of systems.
- **Evaluate** proclaimed advancements in human biology.

This human biology book gives students a chance to explore how the human body works. You will focus on bones, muscles, nerves and hormones, heart and blood vessels, lungs, digestive organs, and kidneys, and how they all work together for "balance" and health, as well as the role of males and females in the process of reproduction. The knowledge you gain from this book will serve you well in many aspects of your life. Your understanding of human biology can benefit your own health. Familiarity with the human body can help you make healthful choices. Your knowledge of human biology can help you understand news about nutrition, medications, medical devices, and procedures and may even help you understand genetic or infectious diseases. Virtually everyone will have some problem with an aspect of their body and knowledge of human biology just might help.

Furthermore, in this human biology book, you will learn about yourself, relating the structure of the different body systems to their function and understanding the interdependence of these systems in maintaining life. You will explore aspects of life, such as coordination of the musculoskeletal, nervous, and endocrine systems, and various responses of the human immune system. In life, you may need to be able to evaluate risks, ethical concerns, and benefits to make informed decisions about matters relating to lifestyle and health. Issues such as diet, medical treatments, and the manipulation of fertility are examples in which personal choices can have an impact on your health and well-being, as are obesity, effects of drugs and alcohol use during pregnancy, and hygiene. These are just a few aspects of human biology of possible concern in your life. Without this understanding, you may make bad health decisions, like buying into the latest fad <u>diets</u>. With an understanding of human biology, you are more likely to make better life decisions, and to be a more effective contributor to discussions related to health issues.

Some fascinating information that people can learn about themselves includes the following:

- The Legend of Vampires: Pellagra, Corn and Niacin Deficiency

- Blood

- How long can people live?

- Where did the Zika Virus come from?

- How does osmosis keep you healthy?

- How do allergies work?

- What happens when you get a sunburn?

- Gluten and You

The human body is so fascinating; it may be overwhelming at first when you start learning about its structure and function. Over time, though, you should appreciate how different parts of the body interact with each other. You will build this understanding slowly, and then find it easier to develop such knowledge. The aims of this book are to explain the human body by exploring the structure and function of cells and tissues, followed by discovering features of the human skeletal, muscular, nervous, endocrine, circulatory, respiratory, digestive, excretory, and reproductive systems.

Homeostasis— The Goal of Human Biology

LESSON 1

Objectives

After completing this lesson, students should be able to

- **define** homeostasis,
- **provide an example** of homeostasis,
- **name** the major body systems,
- **explain** the function of each system, and
- **distinguish between** positive and negative feedback mechanisms.

Defining Homeostasis

Homeostasis, or balance, is the maintenance of the internal environment.

The term homeostasis was first described by Walter Cannon, an American physiologist (Cannon, 1929) as the physical and chemical properties that an organism must maintain to allow proper functioning of its cells, tissues, organs, and organ systems. You have most of your cells protected from the external environment. They are surrounded by a watery internal environment. This internal environment must be maintained to allow maximum efficiency. The ultimate control of homeostasis, or balance, is done by the coordination of many of the body's systems. Often this control is in the form of negative feedback. Your thermostat is a common example of homeostasis that involves the integration of skin, muscular, nervous, and circulatory systems. You have many organs and organ systems that function in homeostasis. Changes in the external environment can trigger changes in the internal environment as a response.

An Example of Homeostasis

The body tends to regulate its temperature via negative feedback (when something increases, it must be decreased, and when something decreases, it must be increased). If body temperature increases, this is noted by the hypothalamus in the brain. Signals are sent to the blood vessels to increase their diameter (vasodilation) and signals also

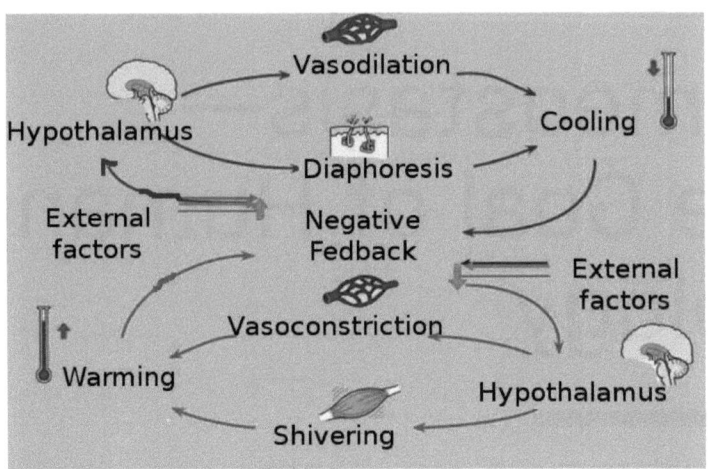

FIGURE 1.1. Negative Feedback

go to sweat glands to initiate sweating (diaphoresis), both of which result in cooling (decreasing of the body temperature). If body temperature decreases, this is noted by the hypothalamus in the brain. Signals are sent to blood vessels to decrease their diameter (vasoconstriction) and signals also go to skeletal muscles to initiate shivering to generate heat, both of which result in warming (increasing of the body temperature). Figure 1.1 shows a simple, familiar example of homeostasis.

The internal environment, such as the components listed below, all require this homeostatic balance:

- concentration of oxygen (O_2) and carbon dioxide (CO_2)
- pH
- nutrients and waste products
- salt and other ions (electrolytes)
- volume and pressure of body fluids, including water balance

Individual Body Systems, Their Functions, and Homeostasis

The body is composed of eleven primary systems.

The skin (integumentary system) is the outer protective layer of the body that prevents water loss and guards against invasion by foreign bacteria and viruses. Sweat glands, hair follicles, and oil glands are located within the skin, as is adipose tissue, which serves as a source of insulation. Nerves, arteries, and veins all run through the skin.

The skeletal system provides support and protection, as well as attachments for muscles. the skeletal system provides a frame upon which movement can occur. It supports and protects the body parts, produces blood cells, and stores minerals, like calcium and phosphorus.

The muscular system allows for movement. The muscular system produces movement and heat, maintains posture, and gives support. The muscular system is closely linked with the skeletal system.

The nervous system coordinates and controls many of the body systems. Memory, learning, and conscious thought are managed by the nervous system. Automatic functions like heartbeat, breathing, and involuntary muscle actions are also performed by the nervous system.

The endocrine system works with the nervous system by secreting hormones to regulate metabolism, growth, and reproduction via chemical communication using the circulatory system.

The circulatory system carries oxygen, carbon dioxide, nutrients, waste products, immune substances, and hormones throughout the body. Circulatory organs include the heart and blood vessels.

The respiratory system exchanges oxygen and carbon dioxide between cells and the blood, as well as helping the kidneys to maintain pH of the blood.

The digestive system breaks down and absorbs food, and eliminates solid wastes. Digestion uses mechanical and chemical processes, breaking down food into absorbable particles, small enough to enter the blood. Absorption of these small food particles happens in the intestines, which allows them to enter the circulatory system.

The renal system manages the amount of body fluids and removes wastes from the body (waste products from the blood, such as urea) in the urine.

The reproductive system, controlled mostly by the endocrine system, is responsible for perpetuating the species. Parts of the reproductive system produce hormones to start, stop, and control sexual development. The reproductive organs produce gametes (eggs and sperm) that combine to generate the next generation, as embryos develop into a fetus and finally a new human.

The immune system defends the body from invading bacteria and viruses, as well as cancer. It also provides cells that help protect the body from disease using an antigen/antibody response mechanism.

Negative and Positive Feedback Systems and Homeostasis

Most of the body's systems use positive or negative feedback to maintain the body's internal environment. Most homeostatic systems are controlled by the nervous and endocrine systems.

The nervous system depends on input from the skin or sensory organs, which send a message to the spinal cord or brain. This sensory input is processed, and a signal is sent to an output system, such as muscles or glands, allowing for a response to the stimulus.

The endocrine system is a second type of control. This involves a chemical change in the body that will send a message to an endocrine gland. For example, insulin is released into the blood when blood glucose (sugar) levels are high. Cells will then remove glucose from the blood, thus lowering the blood glucose levels. This will then shut down the production and release of insulin.

Another Example of Negative Feedback

The body regulates calcium levels via negative feedback (when Ca^{++} increases, it must be decreased, and when Ca^{++} decreases, it must be increased). If Ca^{++} increases, the thyroid gland releases the hormone calcitonin, which prevents bone destruction by osteoclasts and also decreases Ca^{++} reabsorption by the kidneys, both of which cause Ca^{++} levels to decrease. If Ca^{++} decreases, the parathyroid gland releases the hormone PTH, which causes bone destruction by osteoclasts and also increases Ca^{++} reabsorption by the kidneys as well as increasing Ca^{++} absorption in the small intestine (with the help of vitamin D), both of which cause Ca^{++} levels to increase.

Normal childbirth is driven by a positive feedback loop. A positive feedback loop results in a change in the body's status, rather than a return to homeostasis. Figure 1.2 shows an example of positive feedback in childbirth: When the head of the fetus pushes up against the cervix (1) it stimulates a nerve signal from the cervix to the brain (2). When the brain receives this signal, it "tells" the pituitary gland to release a hormone called oxytocin (3). Oxytocin is then carried via the bloodstream to the uterus (4) where it causes contractions (5). This pushes the fetus towards the cervix. This cycle is repeated, eventually inducing childbirth.

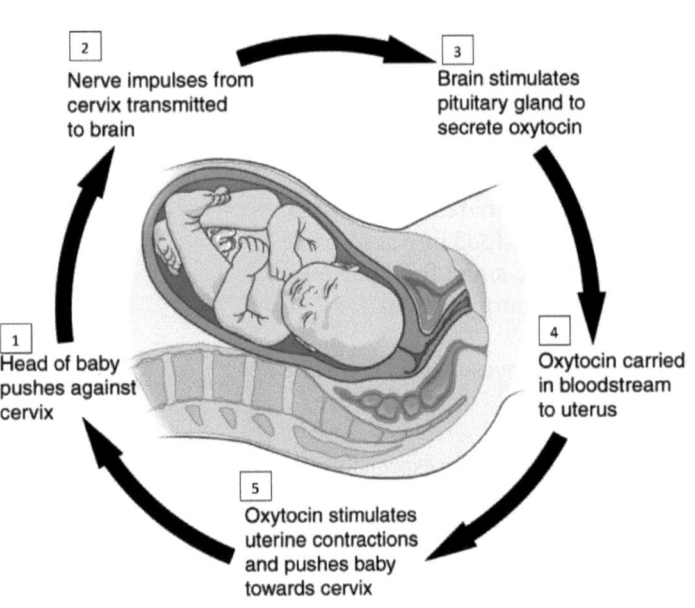

FIGURE 1.2. Positive Feedback

Key Points

Interstitial fluid is the fluid found in the spaces between cells. It bathes and surrounds the cells of the body, and provides a means of delivering materials to the cells and removing metabolic waste.

Homeostasis is the body's ability to maintain its internal environment through adjustments.

- The internal environment involves the blood and interstitial fluid.
  - Blood delivers oxygen and nutrients to the tissues and carries carbon dioxide and wastes away.
  - Interstitial fluid is where substances are exchanged between cells and blood.

- The nervous and endocrine systems are particularly important in maintaining homeostasis.

    - The nervous system issues commands to effector organs.
    - Endocrine glands release hormones that travel through the blood to reach their targets.
    - All of the organ systems work together to maintain life and health.

- Control for your body temperature is located in a part of the brain called the **hypothalamus**.

    - When body temperature is above normal, the hypothalamus "tells" the skin blood vessels to dilate. More blood flow near the surface allows heat to be lost to the environment. The nervous system also activates sweat glands. Evaporation of sweat helps lower body temperature.
    - When body temperature falls below normal, the hypothalamus directs the skin blood vessels to constrict, preserving heat. It also sends nerve signals to skeletal muscles. Shivering generates heat.

- Negative feedback is a homeostatic mechanism to maintain balance.

    - A home thermostat can be used to demonstrate the negative feedback mechanism. The thermostat has a set point. When the room temperature is above or below the set point, it turns the furnace off and on to maintain the preset temperature.

- Positive feedback causes an increasing change in the same direction.

    - A common example of positive feedback involves release of oxytocin during childbirth.

> Effector organs can be muscles that contract or glands that secrete substances in response to nerve signals.

> The hypothalamus is a part of the brain that coordinates both the autonomic nervous system and the activity of the pituitary gland, to control body temperature, thirst, hunger, and other homeostatic systems. It is also involved in sleep and emotional activity.

Reference

W.B. Cannon. "Organization for Physiological Homeostasis," *Physiol.* Rev., 9 (1929): 399–431.

Figure Credits

Fig. 1.1: Source: https://commons.wikimedia.org/wiki/File:Temperature_Regulation.jpg.

Fig. 1.2: Copyright © by Rice University (CC BY 4.0) at https://opentextbc.ca/anatomyandphysiology/chapter/1-5-homeostasis/.

The Mysteries of Science—Scientific Method versus Serendipity

LESSON 2

Objectives

- **Describe** the general process of the scientific method.
- **Differentiate** science from serendipity.
- **Provide** an example of each.

Science, the Scientific Method, and Serendipity

Science involves special skills such as investigational and experimental techniques. Understanding science is important so that when you come across something deemed

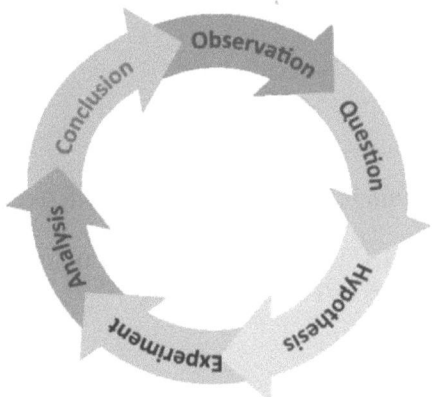

FIGURE 2.1. The Scientific Method

to be scientific in your life, you will be able to apply these investigational and experimental scientific concepts and theories. Serendipity may be considered an "accidental" discovery. In science it may involve finding one thing while looking for something else. Serendipity has played an important role in the science of both biology and medicine.

The **scientific method** (Figure 2.1) is a standardized sequence of events to test an idea regarding what was observed in nature. It involves this sequence of events:

1. Make an observation of something in nature.
2. Question what could have caused what you observed.
3. Form a hypothesis of what you think caused what you observed.
4. Experiment to test your hypothesis.
5. Analyze the experimental results.
6. Draw conclusions as to whether or not your hypothesis was proven correct.

Serendipity means an unplanned, fortunate discovery. The notion of serendipity has existed throughout the history of scientific innovation (Oxford Dictionary). Examples of serendipity include Alexander Fleming's accidental discovery of penicillin in 1928, the invention of the microwave oven by Percy Spencer in 1945, the invention of the Post-it® note by Spencer Silver in 1968, and many more.

> Hypothesis: An idea that is suggested as a possible explanation for a particular situation or condition, but which has not yet been proved to be correct.

Key Points

Science is a way of knowing about the world. This usually involves the scientific method—observation, hypotheses, experiments, conclusions, and maybe new hypotheses.

- After making observations and gathering knowledge, a scientist uses inductive reasoning to form a hypothesis. A hypothesis is based on informed knowledge (not a guess). It can then be tested by getting more data through experimentation.
- An experiment is used to test a hypothesis.
- Analysis allows for finding relationships in the results of the experiment.
- Finally, the data is analyzed to reach a conclusion—is the hypothesis supported or not?

Serendipity can be defined as an accidental discovery. In science it may be defined as finding one thing while looking for something else. It is interesting that serendipity has played an important role in the science of biology and medicine.

- Some additional examples of serendipity include the discovery of heparin, Dramamine®, X-rays, gram stains, the pancreas's role in diabetes, anesthetics like ether and nitrous oxide, some artificial sweeteners, LSD, Viagra®, and many others.

Reference

OxfordDictionaries.com. *Oxford Dictionary*. Retrieved 23 April 2018.

LESSON 3

The Mysterious Bond between Chemistry and Human Biology

Objectives

- **Describe** the structure of an atom.
- **Distinguish** between ionic and covalent bonds.
- **Explain** the role of hydrogen bonds in the properties of water.
- **List** the four classes of organic molecules that are found in cells.
- **Compare** the structures and functions of fats, phospholipids, and steroids.
- **Describe** the structure of an amino acid.
- **Explain** the difference between RNA and DNA.

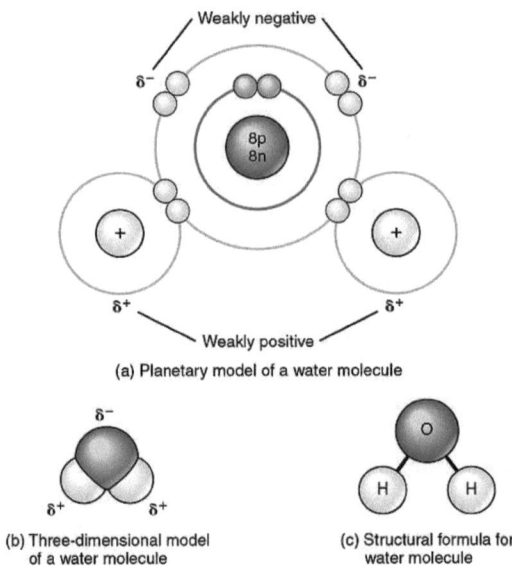

FIGURE 3.1. The Water Molecule

Figure 3.1. The Water Molecule shows the "magic" of the structure of water that allows for life, as we know it, to exist. Note the unequal sharing of electrons between the hydrogen atoms and the oxygen atom. These weak hydrogen bonds give water most of its unique properties. **Polar molecules** (like water) unequally share electrons between atoms, so have a slight positive charge at one end and a slight negative charge at the other.

Atoms

Figure 3.2 shows the periodic table of elements. Each element has a name and symbol, atomic number (number of protons), and atomic mass.

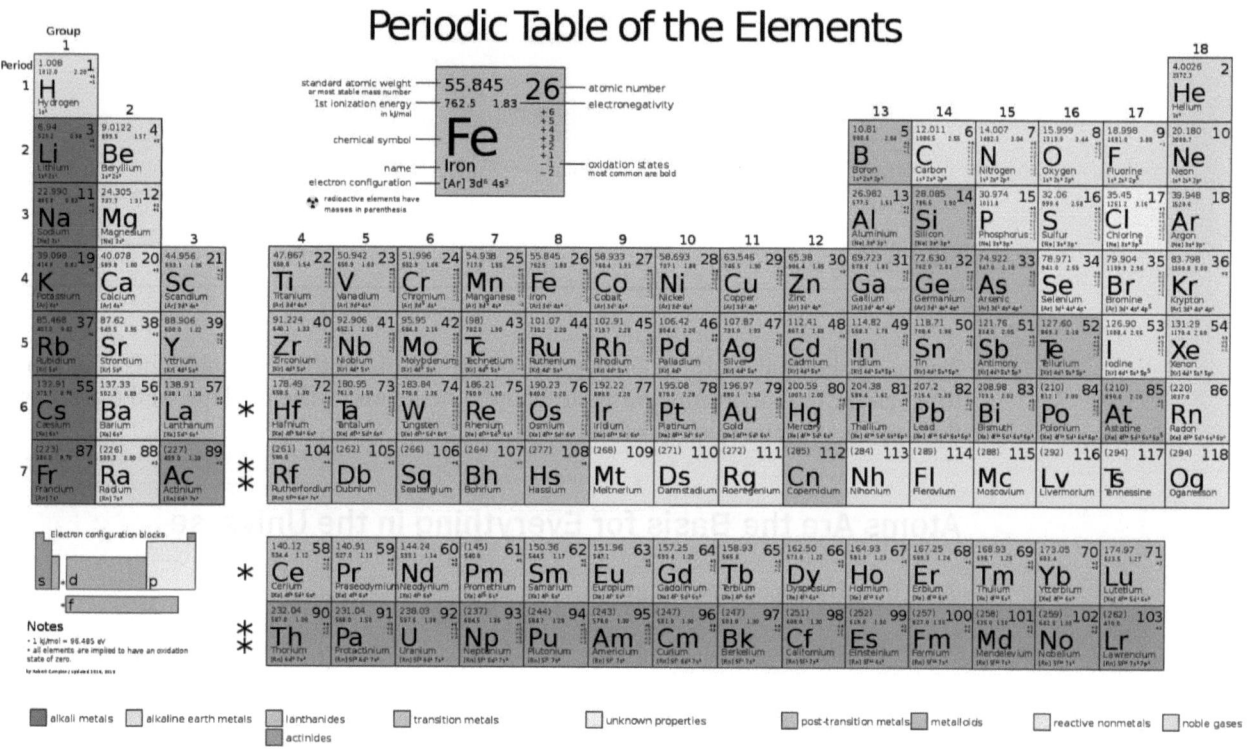

FIGURE 3.2. The Periodic Table

Elements: substances that can't be broken down any further

Atom: the smallest unit of an element

Chemical symbol: begins with **one or two letters** based on the element's name

Two or more atoms joined together chemically: molecule

Molecule containing at least two different elements: compound

LESSON 3 The Mysterious Bond between Chemistry and Human Biology | 11

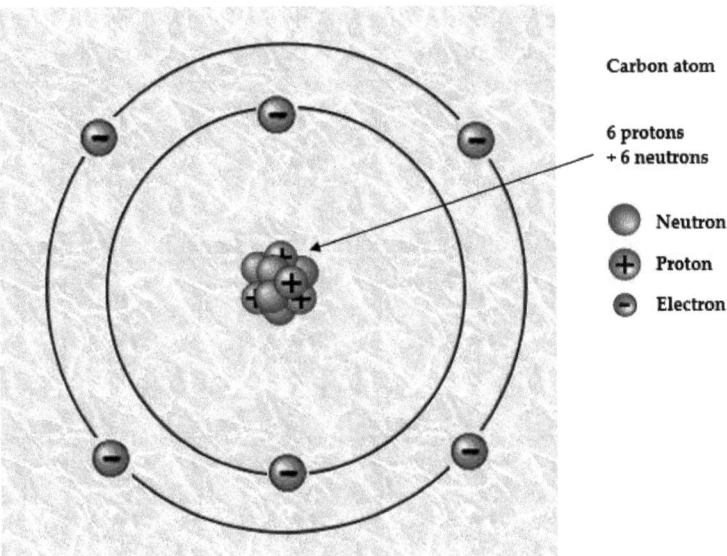

FIGURE 3.3. A Carbon Atom

Figure 3.3 shows an example of an atom, using carbon for this example. There are 6 protons (giving carbon atomic number 6) and 6 neutrons in the nucleus, and 6 electrons floating around the nucleus of the atom.

Atoms Are the Basis for Everything in the Universe

The three basic parts of an atom:

- **Protons** = + positive charge

    - Part of the atomic nucleus
    - Repel each other

- **Neutrons** = neutral (a charge of zero)

    - Part of the atomic nucleus
    - Separate protons, making an atom more stable

- **Electrons** = - negative charge

    - Orbit nucleus in different shells, or energy levels

The thing that makes each element unique is the number of protons, since the number of neutrons and electrons can vary.

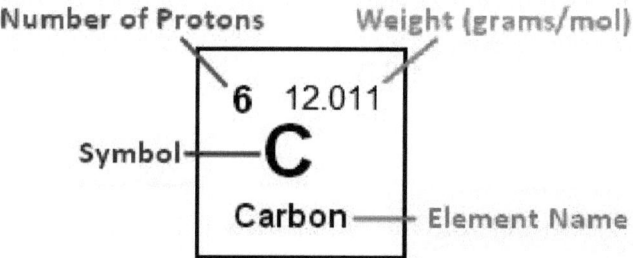

FIGURE 3.4. Carbon Example from Periodic Table

Figure 3.4 shows an example of an entry in the periodic table, using carbon. It shows carbon's symbol (C), name (carbon), number of protons (6, carbon's atomic number), and atomic weight (12.011).

- **Atomic number:** the **number of protons** in the nucleus of an atom
- **Atomic mass** (aka atomic weight): The atomic mass of an element is rarely an even number. This happens because of the **isotopes.**

    o Many elements occur as **isotopes**. They vary in the number of **neutrons** they have.

- **Mass number:** the number of protons, plus the number of neutrons
- In a neutral atom, there are the same number of protons (+) and electrons (−).

Radioactive Atoms

- Isotope is **radioactive** if nucleus is unstable.
- Most isotopes disintegrate spontaneously with the release of energy by processes of **nuclear** or **radioactive decay**.
- When the nucleus changes in structure, energy and/or subatomic particles are given off.

Electrons

- Electrons orbit around the atomic nucleus in **shells**.
- The inner shell of an atom, closest to the nucleus, can have a maximum of two electrons.
- The outermost shell is called the valence shell.
- Eight (8) is the *maximum number* of valence electrons for a full valence shell.
- Number of valence electrons governs an atom's bonding behavior.
- Atoms are much more stable, or less reactive, with a full valence shell.

Bonding

- Stability can be achieved one of two ways:
    - **ionic** bond
    - **covalent** bond

- By moving electrons, the two atoms become linked in what is known as **chemical bonding**.

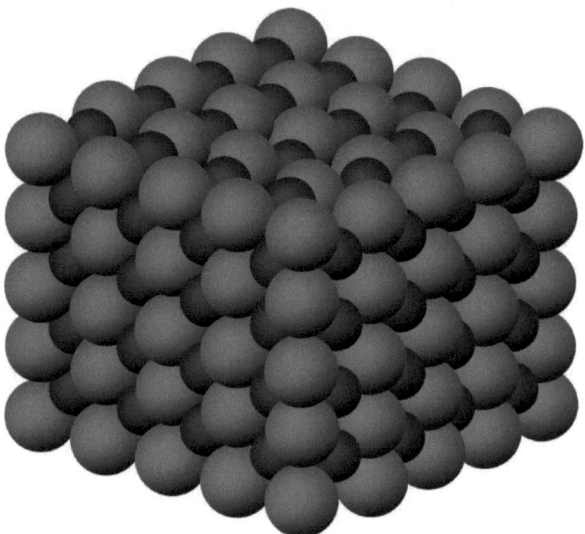

FIGURE 3.5. Ionic Bonds

Ionic Bonds

- Involves transfer of electrons between two atoms
- **An ion** is an atom or group of atoms that has lost or gained one or more electrons, making it negatively or positively charged.
- Ionic bonds are atoms held together by attraction between a (+) **and a** (−) **ion.**
- **Compound is neutral overall,** but still charged on the inside
- Makes solid crystals

Covalent Bonds

- Involves the sharing of a pair of electrons between atoms
- One covalent bond = 1 pair of shared electrons
- Covalent compounds can make single (2 electrons), double (4 electrons), or even triple bonds (6 electrons), depending on the number of electrons they share.
- **Found mainly in nature as organic compounds**

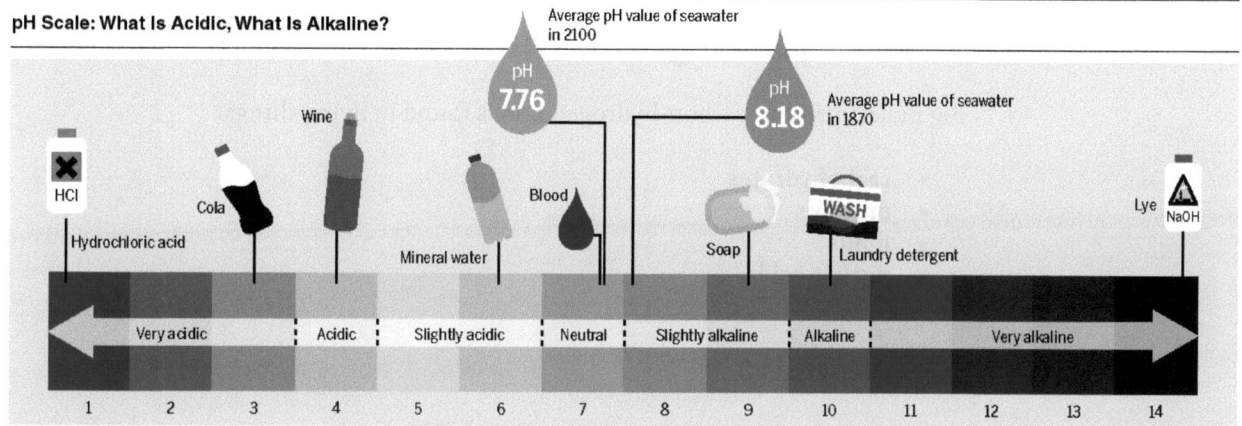

FIGURE 3.6. pH

Figure 3.6 shows the pH scale with everyday examples. pH is a measure of the number of hydrogen ions (H⁺) in a solution.

Acids and Bases

- An **acid** is any ionic compound that releases **hydrogen (H⁺)** in solution.
- A **base** is any ionic compound that releases **hydroxide (OH⁻)** in solution.
- **Acidity of a solution** is measured by concentration of hydrogen ions (H⁺) versus hydroxyl ions (OH⁻).
- pH balance is important to homeostasis of organisms.

Examples:

- Digestion, occurring in the stomach, needs an acidic environment (pH 2–3).
- Urine is slightly acidic.
- Blood must stay in neutral range near 7.35 to 7.45.

Buffers
Buffers (substances that resist a change in pH) can combine with excess hydrogen (H+) or hydroxide (OH-) ions.

- Produce substances less acidic or alkaline
- Act like a chemical sponge to soak up excess acid or base, keep pH constant
- Buffers are vital to maintaining pH in organisms.

The Major Organic Macromolecules

Big molecules with carbon-hydrogen bonds found in living things:

- carbohydrates
- proteins
- lipids
- nucleic acids

Carbohydrates

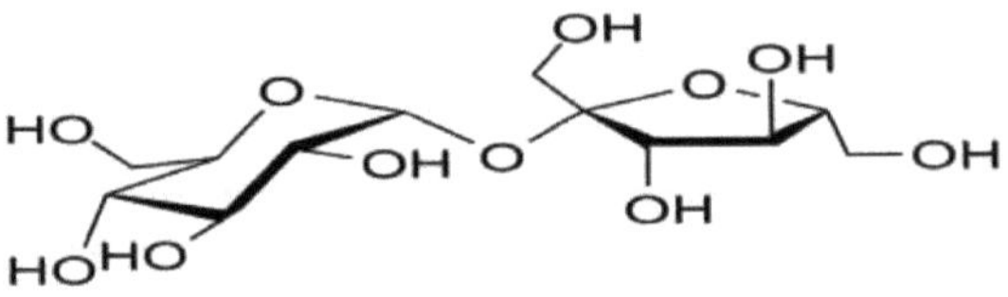

FIGURE 3.7. A Disaccharide

Figure 3.7 is a disaccharide, an example of a carbohydrate. This is sucrose, table sugar, made of the two simple sugars glucose and fructose held together by a glycosidic bond, which can be broken by the enzyme sucrase.

- "carbon"-hydrates
- One carbon molecule to one water molecule $(CH_2O)_n$
- **Saccharide** is a synonym for carbohydrate.
- The prefixes on the word saccharide can be mono-, di-, tri-, or poly- (based on size).

Enzymes: Biological molecules that can speed up a chemical reaction—always end with the suffix -ase.

Proteins

Figure 3.8 shows the basic structure of an amino acid (a central carbon with an amino group, an acid group, a hydrogen, and an R group, which differentiates the 20 naturally

Monomers and polymers: Monomers are single units (simple sugars for carbohydrates, amino acids for proteins); polymers are multiple monomers combined to make larger molecules.

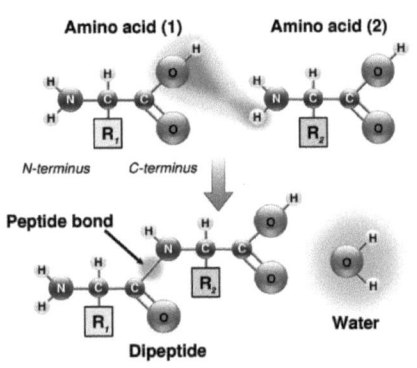

FIGURE 3.8. Amino Acids

occurring amino acids). Also shown is a dipeptide—two amino acids held together by a peptide bond.

Proteins are macromolecules—**polymers** composed of monomers called amino acids. **Amino acids** contain a

- base amino group (–NH$_2$)
- acidic carboxyl group (–COOH)
- hydrogen atom

... all attached to same carbon atom (the α-carbon [alpha carbon]).
A fourth bond attaches the α-carbon to a side group (-R) that varies among amino acids. The side group (-R) is important—it affects the way a protein's amino acids interact with one another, and how a protein interacts with other molecules.

Essential amino acids cannot be synthesized by the body. They must be ingested in the diet.

Lipids

FIGURE 3.9. Examples of Various Lipids (Fats), and an Image Showing That Water and Fat Don't Mix

Figure 3.9 shows various lipids (fats): Cholesterol, at the top, is necessary for all cell membranes, and is the building block for substances like steroid hormones; a free fatty acid, which can be saturated or unsaturated (as in this image), can be used by the body for energy (ATP), or attached to a glycerol molecule forming a triglyceride—the storage form of fat in the body; a triglyceride, the storage form of fats in the body, which can provide a lot of energy when broken down; and a phospholipid, a special triglyceride-like lipid, which is the main component of all cell membranes (phospholipid bilayers).

Fats

Fats and oils are made from two kinds of molecules:

- **glycerol** (a type of alcohol)
- **fatty acids** (with glycerol form triglycerides)

Phospholipids

Phospholipids are a major component of all cell membranes.

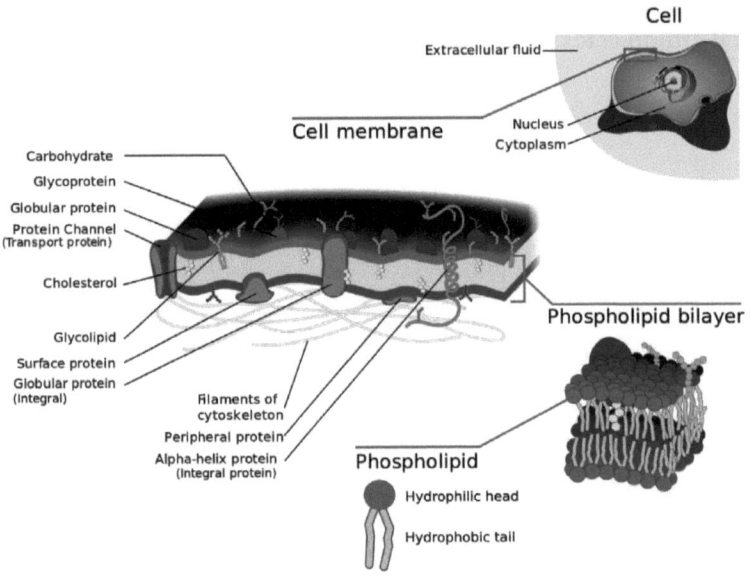

FIGURE 3.10. The Importance of Phospholipids in the Cell Membrane

Figure 3.10 shows many components of the cell that use lipids, especially phospholipids.

- Most phospholipids contain a diglyceride as the tail, and a phosphate group for head.
- Hydrocarbon tails are **hydrophobic**, but phosphate heads are **hydrophilic**.
    - So phospholipids are soluble in both water and oil.

With non-polar/hydrophobic tails from both layers facing inward and the polar/hydrophilic heads facing outward a **phospholipid bilayer** is created (as shown in Figure 3.10).

Steroids

- The central core of a cholesterol molecule (4 fused rings) is shared by all steroids.
- Cholesterol is precursor to our **sex** hormones and Vitamin **D**.

- Our cell membranes contain cholesterol (in between the phospholipids) to help keep the membrane "fluid."

Nucleic Acids

Hydrophilic versus hydrophobic: Water loving versus water fearing. Polar substances (charged particles) love water, whereas nonpolar substances "fear" water (fats).

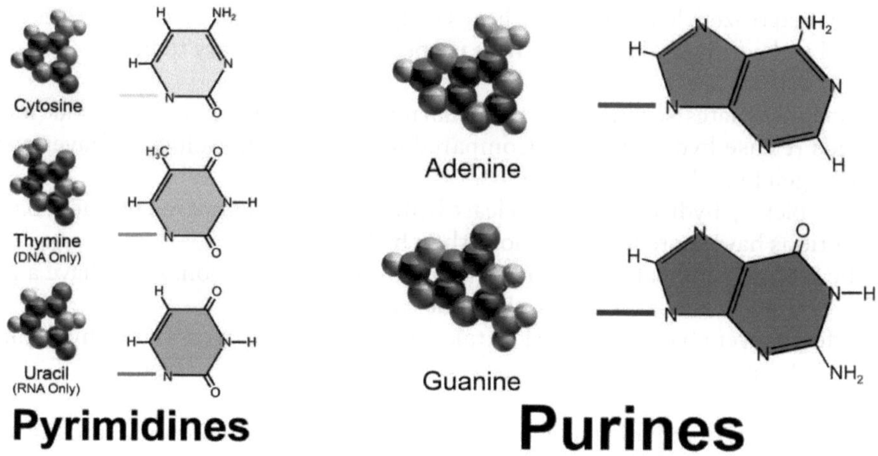

FIGURE 3.11. The Nucleic Acids—Pyrimidines and Purines

Figure 3.11 shows the various nucleic acids that, primarily, make up DNA and RNA in each cell. Adenine is also a key part of ATP (Adenosine TriPhosphate).

Nucleic Acid Structure (DNA)

- Nucleotides are linked by covalent bonds between the **sugar** of one nucleotide and the **phosphate** of next (**sugar-phosphate backbone**).
- Nitrogenous **bases** extend from it like rungs of a ladder.

Key Points

- All matter is composed of elements. Every element has a name and symbol.
- Atoms contain subatomic particles. Positively charged protons and neutral neutrons are in the nucleus of the atom. Negatively charged electrons orbit around the nucleus. The atomic number equals the number of protons. Atomic weight equals the number of protons plus neutrons.
- Isotopes are atoms that have the same atomic number but differ in the number of neutrons.
- Atoms bond with each other to form molecules, or compounds.
- Ionic bonds involve atoms that either give up or receive electrons to achieve stability.
- Covalent bonds involve sharing of pairs of electrons to achieve stability.

- Life as we know it would be impossible without water and its hydrogen bonds. Water is polar: the oxygen has a slight negative charge and the hydrogen has a slight positive charge.
- Hydrogen bonds are formed when a slightly positively charged hydrogen atom is attracted to a negatively charged atom.

Due to being polar and its hydrogen bonding, water is a liquid at room temperature, is less dense when frozen, loses and gains heat slowly, has a high heat of vaporization, fills vessels, and is the universal solvent. These properties are necessary to life.

> **Heat of vaporization:** The amount of heat required to convert a liquid into a gas at constant temperature and pressure.
>
> **Solvent:** A molecule that has the ability to dissolve other molecules, known as solutes.

- Water dissociates (separates) into an equal number of hydrogen and hydroxide ions.
- Acids release hydrogen ions. Compared to water, acidic solutions have more hydrogen ions than hydroxide ions.
- Bases pick up hydrogen ions or release hydroxide ions. Compared to water, basic solutions have more hydroxide ions than hydrogen ions.
- The pH scale indicates the acidity or alkalinity of a solution. Acids have a pH lower than 7, and bases have a pH higher than 7.
- Buffers prevent a change in pH by taking up excess hydrogen ions or hydroxide ions. Maintaining pH within a narrow range is important to health.

The four categories of organic molecules—carbohydrates, lipids, proteins, and nucleic acids—are the molecules of life. Macromolecules are formed when monomers join together to form polymers.

Carbohydrates provide an energy source for cells.

- The monosaccharide (simple sugar) glucose provides a ready source of energy for cells.
- Lactose is the disaccharide (two monosaccharides—glucose and galactose) found in milk.
- Polysaccharides are polymers of glucose (starch and glycogen in plants and animals, respectively; cellulose, a polysaccharide found in plant cell walls, is referred to as fiber because the bonds in cellulose cannot be digested).

Proteins are extremely important for the structure and function of cells. Their functions include support, enzymes, transport, defense, hormones, and motion.

- An amino acid has a central carbon with a hydrogen connected, and it also contains an amino group, an acid group, and an *R* (remainder) group to distinguish the 20 different amino acids from one another.
- Amino acids are joined by a peptide bond. Three or four amino acids bonded together are called a polypeptide.
- When proteins are exposed to extremes in heat and pH, they undergo denaturation (drastic change in structure). This destroys the proteins' function.
- All proteins can be described by looking at their primary, secondary, tertiary, or quaternary structures. The primary structure of a polypeptide is the sequence of amino acids. The secondary structure is an alpha helix or pleated sheet. The tertiary structure typically is globular. Quaternary structure, if it exists for a protein, occurs if there is more than one polypeptide chain (subunit, like hemoglobin).

Lipids are hydrophobic and do not mix with water (e.g., Italian salad dressing). They function in the storage of energy, structure of cell membranes, and steroid hormones.

- Solid fats of animal origin and liquid oils of plant origin are triglycerides. Emulsifiers are substances that can allow fat droplets to mix with water (like soap!).
- Saturated fatty acids have no double bonds, and unsaturated fatty acids do have double bonds. Saturated fats are associated with cardiovascular disease. The most harmful fats, more than any naturally occurring fats, are the trans fats, which are in vegetable oils that have been partially hydrogenated to make them more solid at room temperature (found in baked goods —cookies, crackers, pastries, etc.).
- Phospholipids, which have a polar phosphate group instead of a third fatty acid, are the primary constituent of the plasma membrane (a phospholipid bilayer).
- Steroids are based on the cholesterol molecule (as are the sex and steroid hormones).

The two types of nucleic acids are DNA (deoxyribonucleic acid) and RNA (ribonucleic acid).

- Both DNA and RNA are polymers of nucleotides.
- A nucleotide contains phosphate, a pentose sugar, and a nitrogen-containing base. The bases in DNA are adenine, thymine, cytosine, and guanine (A, T, C, G). In RNA, uracil replaces thymine (A, U, C, G).
- DNA is a double helix resembling a stepladder. Phosphate and a pentose sugar make up the sides of the ladder. Hydrogen-bonded bases are the rungs. RNA is single stranded.
- ATP is a nucleotide modified by the addition of three phosphate groups. It functions as the energy currency in cells. The energy released by ATP breakdown is used by the cells for various functions.

Figure Credits

Fig. 3.1: Copyright © by OpenStax College (CC BY 3.0) at https://commons.wikimedia.org/wiki/File:209_Polar_Covalent_Bonds_in_a_Water_Molecule.jpg.
Fig. 3.2: Copyright © by 2012rc (CC BY 3.0) at https://commons.wikimedia.org/wiki/File:Periodic_table_large.svg.
Fig. 3.3: Copyright © by Alejandro Porto (CC BY-SA 3.0) at https://commons.wikimedia.org/wiki/File:Carbon-atom.jpg.
Fig. 3.5: Source: https://pixabay.com/vectors/crystal-structure-nacl-chemical-148812/.
Fig. 3.6: Copyright © by Heinrich-Boll-Stiftung (CC BY-SA 2.0) at https://commons.wikimedia.org/wiki/File:PH_Scale-_Acidic_vs._Basic_(Alkaline).png.
Fig. 3.7: Source: https://chemistry.stackexchange.com/questions/15442/how-does-sulfuric-acid-dehydrate-sugars.
Fig. 3.8: Source: https://commons.wikimedia.org/wiki/File:Peptidformationball.svg.
Fig. 3.9a: Copyright © by Eoin Fahy (CC BY-SA 3.0) at https://commons.wikimedia.org/wiki/File:Common_lipids_lmaps.png.
Fig. 3.9b: Copyright © by Bitjungle (CC BY-SA 2.0) at https://commons.wikimedia.org/wiki/File:Oil_in_water.jpg.
Fig. 3.10: Copyright © by Dhatfield (CC BY-SA 3.0) at https://commons.wikimedia.org/wiki/File:Cell_membrane_detailed_diagram_3.svg.

Fig. 3.11a: Copyright © by BruceBlaus (CC BY 3.0) at https://commons.wikimedia.org/wiki/File:Blausen_0324_DNA_Pyrimidines.png.

Fig. 3.11b: Copyright © by BruceBlaus (CC BY 3.0) at https://commons.wikimedia.org/wiki/File:Blausen_0323_DNA_Purines.png.

The Mysteries of Structure
Cells

LESSON 4

Objectives

- **List** the organelles of a typical eukaryotic cell.
- **Describe** the structure of the plasma membrane.
- **List** the type of molecules found in the plasma membrane.
- **Identify** the role of the plasma membrane and the various organelles in a human cell.
- **Distinguish** among diffusion, osmosis, and facilitated transport, and state the role of each in the cell.
- **Compare** passive-transport and active-transport mechanisms.
- **Explain** the importance of the mitochondria in cellular energy production.
- **Identify** the role of an enzyme in a metabolic reaction.
- **Summarize** the roles of anaerobic and aerobic pathways in energy generation.
- **Explain** the role and location of the ribosomes.
- **Describe** the structure and function of the organelles of the endomembrane system.
- **Describe** the structure of the nucleus and explain its role as the storage place of genetic information.
- **Discuss** the principles of Mendelian genetics.

Many vital processes occur in the cell. Different cells may have different structures and numbers of organelles (tiny, cellular organs) to enable them to perform their specific functions. These cells together form tissues which group together to form organs and organ systems that then work with different cell types to enable systems to perform coordinated functions—all with the goal of homeostasis. There are many basic functions that are common to most cells in the body, along with a range of special functions performed by specific cells or cell types.

Figure 4.1 shows a typical eukaryotic cell and its organelles.

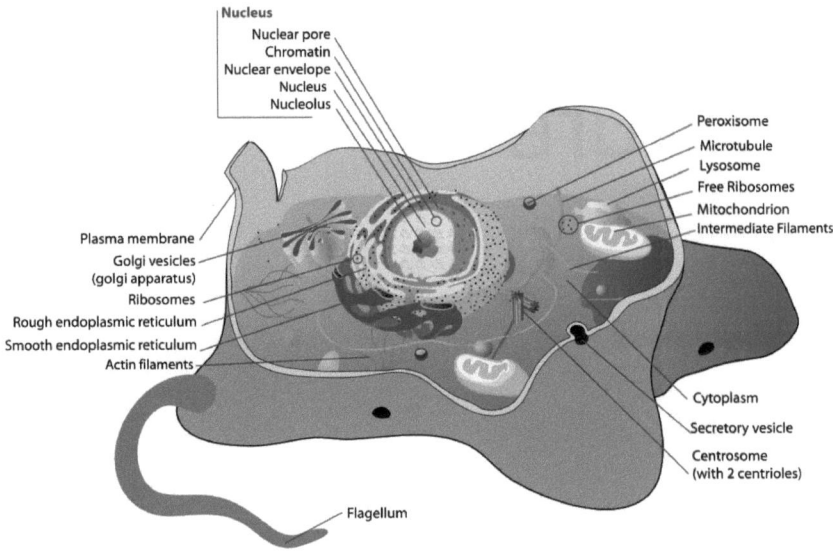

FIGURE 4.1. The Cell

Cells

The Cell Theory

- All living things are composed of cells;
- chemical reactions take place inside cells;
- all cells originate from other cells; and
- cells contain hereditary information that is passed from one generation to the next.

Cells and Cellular Organization

- Cells are the building blocks of life!
- All living things are made of one or more cells.
- Cells only come from other cells.

Life often demonstrates organization. Atoms are organized into molecules; molecules and compounds are organized into cellular organelles, and organelles are organized into cells, which then form tissues, organs, organ systems, and finally organisms. All living things are composed of cells. Prokaryotes (pro = before, karyo = nucleus) may have been the first form of life on Earth (like bacteria). Eukaryotes (eu = true, karyo = nucleus) make up the majority of other life forms, and have membrane-bound nuclei and other intracellular organelles (like us!). All cells have certain features in common, such as a cell membrane, cytoplasm, ribosomes, and DNA and RNA. Eukaryotic cells contain these many organelles and other structures within.

The Cell Membrane

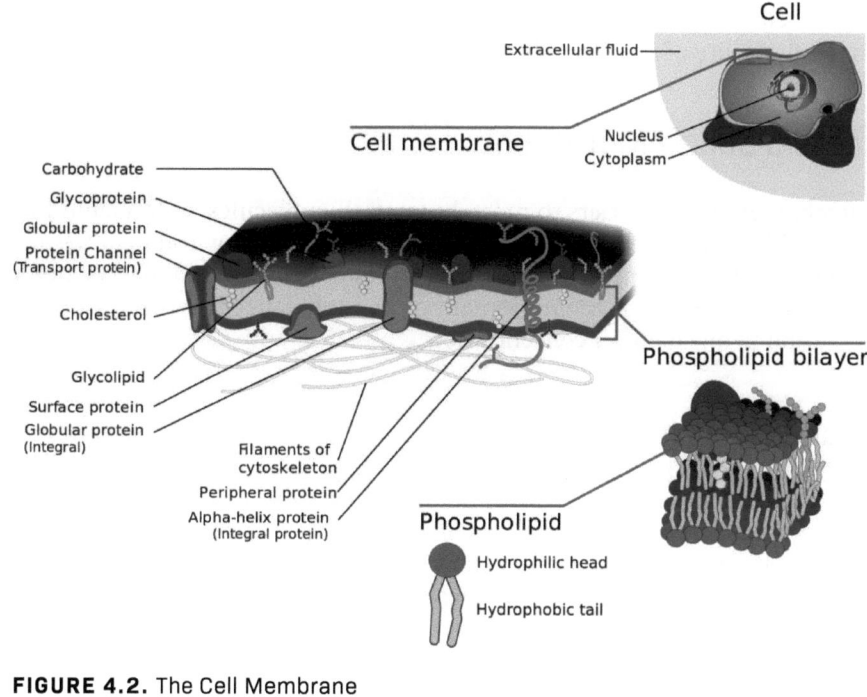

FIGURE 4.2. The Cell Membrane

Figure 4.2 shows the cell membrane (aka plasma membrane or plasmalemma) that surrounds all cells. It

- separates the inner parts of the cell from the outside; and
- is selectively permeable—allows only certain chemicals (like water) through.

The primary function of plasma membrane is to regulate movement of molecules entering or leaving cell.

Some proteins "stick out" from the cell membrane. These proteins may function as channels to help certain molecules cross into and out of the cell. The outer surface of the membrane will tend to contain glycolipids (a combination of carbohydrates and lipids). Many organisms have these structures on the membrane surface to help them recognize "self." Antigens are foreign substances recognized as "non-self." Antibodies are molecules (Y-shaped) that recognize a specific antigen, and can attack them. This is the basis of immunity and vaccination.

Transport into and out of Cells

The cell membrane, as a semipermeable barrier, allows only certain molecules to cross it while keeping most substances from getting inside the cell. It is composed of a phospholipid bilayer. These phospholipids have a polar (hydrophilic) head and two nonpolar (hydrophobic) tails, and are aligned tail to tail so the nonpolar areas form a hydrophobic region in the center of the membrane between the hydrophilic heads, which are on the surfaces of the membrane. Cholesterol is another important component of cell membranes. It is embedded in the hydrophobic areas of the tail to tail region. Cholesterol aids in the flexibility of a cell membrane.

Water and Solute Movement

Cell membranes act as barriers to most, but not all, molecules. Cell membranes are semipermeable barriers separating the inner from the outer environment. Osmosis is the tendency of water to move from an area of higher concentration to one of lower concentration. Diffusion is the movement of any substance from an area of higher concentration to one of lower concentration.

Cells and Diffusion

Water, carbon dioxide, and oxygen are among the few simple molecules that can cross the cell membrane by diffusion (movement of substances from high concentration to low concentration) or osmosis (diffusion of water). Carbon dioxide is produced by all cells as a result of cellular metabolism. Metabolic processes usually require oxygen.

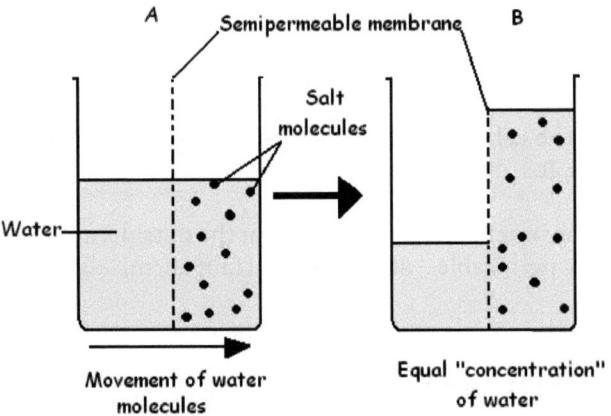

FIGURE 4.3. Diffusion across a Semipermeable Membrane

Figure 4.3 shows **diffusion**, when molecules move down a concentration gradient, from a higher to a lower concentration. **Osmosis** is the diffusion of water.

Tonicity and Osmosis Definitions

- **Isotonic:** equal concentration of a solute inside and outside of cell
- **Hypertonic:** a higher concentration of solute
- **Hypotonic:** a lower concentration of solute

Hypertonic solutions (which produce shrinking of cells, called crenation) are those in which more solute (whatever is dissolved in solvent) is present. Hypotonic solutions (which produce swelling of cells) are those with less solute dissolved in solvent. Isotonic solutions have equal concentrations of substances in solution. One of the major functions of blood in animals is to maintain an isotonic internal environment. This eliminates the problems associated with water loss or excess water gain in or out of cells. Again, we have homeostasis.

Passive, Carrier-Assisted, and Active Transport

Figure 4.4 shows examples of mechanisms of facilitated transport (where proteins assist in diffusion of molecules across the plasma membrane).

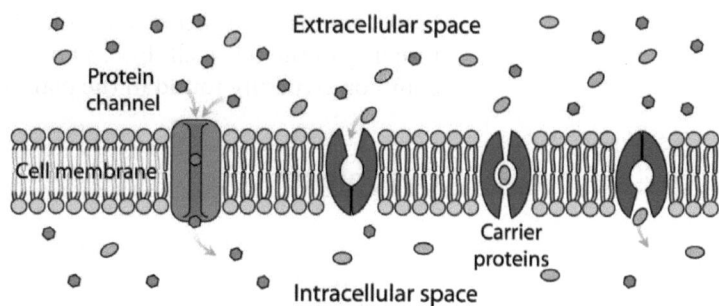

FIGURE 4.4. Facilitated Transport

Passive transport requires no energy from the cell (diffusion of oxygen and carbon dioxide, osmosis of water, and facilitated diffusion). The carrier-assisted transport proteins are integrated into the cell membrane. Some of these proteins can move materials across the membrane when assisted by the concentration gradient (facilitated transport). Active transport requires the cell to use energy (ATP) to accomplish transport. With active transport, the proteins have to move against the concentration gradient (from low concentration to high concentration, like the sodium-potassium pump in nerve cells). Na^+ is maintained at low concentrations inside the cell and K^+ is at higher concentrations. The reverse is true on the outside of the cell. When a nerve signal is sent, the ions pass across the membrane. After the message is sent, the ions must be actively transported back to their "starting positions." About one-third of the ATP used at rest is used to reset the Na-K concentration gradients across the membrane.

Figure 4.5 shows how energy from adenosine triphosphate (ATP) drives substances across the plasma membrane with the aid of carrier molecules.

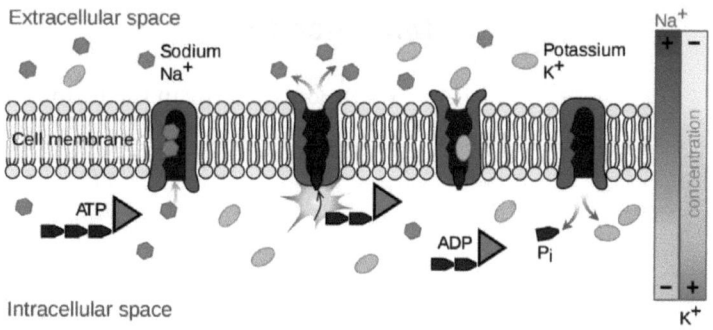

FIGURE 4.5. Active Transport

Vesicle-Mediated Transport

Vesicles and vacuoles (containers) that fuse with the cell membrane may be utilized to release or transport chemicals out of the cell or to allow them to enter a cell. Exocytosis (spitting out) is the term applied when transport is out of the cell. Endocytosis is essentially ingesting a substance into a vesicle (eating). Phagocytosis is the type of endocytosis where an entire cell is engulfed. Pinocytosis is when the external fluid is engulfed (drinking). Receptor-mediated endocytosis occurs when the material to be transported binds to certain receptors on the membrane (the transport of insulin and cholesterol into cells).

Cell Organelles

The Cell Wall

Not all living things have cell walls. The cell wall is located outside the plasma membrane. Animal cells lack a cell wall. They rely on their cell membrane to maintain the integrity of the cell. Cellulose, a nondigestible (to humans) polysaccharide, is the most common structure found in the plant cell wall.

Cytoplasm

- The cytoplasm is the material inside the plasma/cell membrane in which everything "floats." It fills the space between the plasma membrane and the nuclear membrane.
- It is a water-like substance that fills cells.
- It consists of the **cytosol** and **cellular organelles,** except for the cell nucleus.
- **The cytosol** is made up of water, salts, organic molecules, and many enzymes that catalyze reactions.

Mitochondria

- Produce energy in the form of ATP.

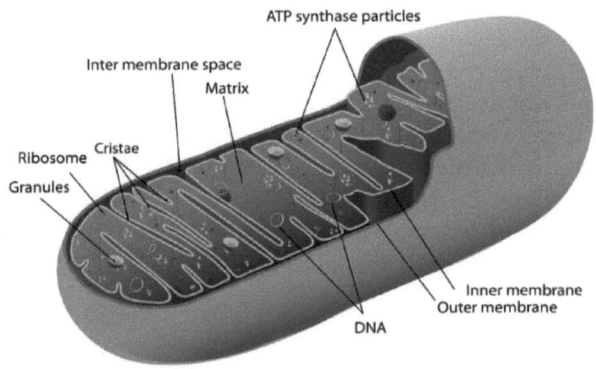

FIGURE 4.6. Mitochondria

Figure 4.6 is a depiction of the mitochondria—the energy factory of the cell. They use various enzymes to produce energy in the form of ATP.

Enzymes

Enzymes are **proteins** that catalyze (speed up) chemical reactions in the body. Enzymes catalyze reactions by weakening chemical bonds, which lowers activation energy (energy needed for a reaction to occur).

- Each enzyme has a unique 3-D shape, including a surface groove called an **active site**.
- The enzyme works by binding a specific chemical reactant (**substrate**) to its active site.
- The resulting **product**(s) is then released from the active site.
- Enzymes are specific for what they will **catalyze**.
- They are not consumed (used up) in the reactions they catalyze.
- Enzymes have names that usually end in **-ase**.

    - **Sucrase**
    - **Lactase**
    - **Maltase**

Factors That Influence Enzyme Activity

1. Temperature

    - Temperatures far above the normal range **denature** enzymes. (This is why very high fevers are so dangerous. They can cook the body's proteins.)

2. pH

    - Most enzymes work best near **neutral** pH (6 to 8).

3. Cofactors and coenzymes

    - **Non-protein** portions of the enzyme (e.g., zinc, iron, copper, vitamins) that are needed for proper enzymatic activity
    - Most **vitamins** are coenzymes essential in helping move atoms between molecules in the formation of carbohydrates, fats, and proteins.

4. Inhibitors—competitive and noncompetitive

    - **Competitive inhibitor**

        - Chemicals that resemble an enzyme's normal substrate and compete with it for the active site
        - May be reversible depending on the concentration of inhibitor and substrate
        - **EXAMPLE:** The drug **Antabuse** is used to help alcoholics quit drinking. Antabuse *inhibits aldehyde oxidase*, resulting in the accumulation of

acetaldehyde during the metabolism of ethanol. Elevated acetaldehyde levels cause symptoms of nausea and vomiting.

- **Noncompetitive inhibitor**

    - Do not enter the active site, but bind to another part of the enzyme, causing the enzyme and active site to change shape
    - Usually reversible, depending on concentration of inhibitor and substrate
    - **EXAMPLE:** You may know that compounds containing **heavy metals** such as lead, mercury, copper, or silver are **poisonous**. This is because ions of these metals are noncompetitive inhibitors for several enzymes.

Metabolism: The Transformation of Energy

- The sum of all chemical reactions in a cell or organism
- Cells either get their energy either by **photosynthesis** or by **eating fuels.**

Metabolic pathways are a series of chemical reactions that regulate the concentration of substances within the organism.

- Have order, like an assembly line
- Molecules are altered in a series of steps.

    - Use many smaller steps rather than one big step

- **Enzymes** are the "workers" that control each step along the pathway.
- May be turned on and off as needed.
- Nearly all chemical reactions in biological cells need enzymes to make the reaction occur fast enough to support life.

Energy is obtained by breaking chemical bonds in foods we eat, like **glucose**.
Metabolism transfers food energy into **ATP energy**, the common energy currency of cells.

- **ATP is Adenosine 5'-triphosphate.**
- ATP is the "molecular currency" of intracellular energy transfer.
- Metabolism releases energy from nutrients.
- That energy can be stored in high-energy phosphate bonds of ATP.
- ATP transports chemical energy within cells.
- ATP can be used to fuel many cellular reactions.

Anabolic Reaction (Anabolism)

The phase of metabolism in which simple substances are **synthesized** into the complex materials of living tissue (building big substances from smaller building blocks).

Catabolic Reaction (Catabolism)

The metabolic **break down** of complex molecules into simpler ones, often resulting in the release of energy.

Carbohydrate Catabolism

- Organisms **catabolize** (break down) **carbohydrates** as the primary energy source for anabolic reactions.
- The monosaccharide **glucose** is most commonly used.
- Glucose is catabolized by

    ○ **anaerobic respiration and fermentation:** only partially breaks down glucose, into pyruvic acid, organic waste products, and a little ATP (Figure 4.7)

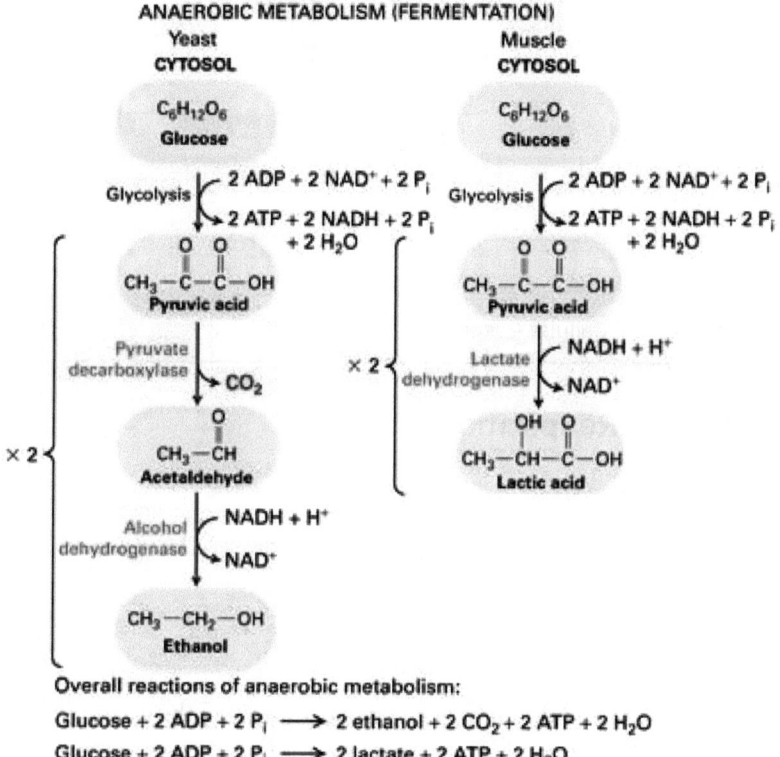

FIGURE 4.7. Anaerobic Respiration and Fermentation

- In fermentation, after glycolysis, there are additional steps to oxidize NADH (into NAD$^+$).
- Electrons and hydrogen ions from the NADH that was produced by glycolysis are donated to another organic molecule.
- No more ATP is created through these additional steps.

- So essentially ...

FERMENTATION = glycolysis + recycling of NAD⁺

Aerobic cellular respiration: results in complete breakdown of glucose to carbon dioxide, water and a lot of ATP

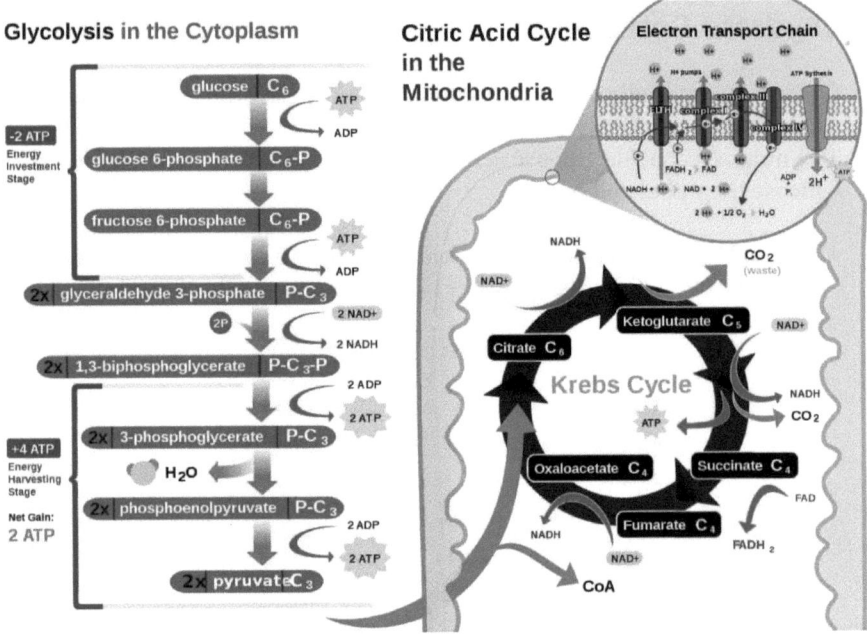

FIGURE 4.8. Aerobic Cellular Respiration

Aerobic Cellular Respiration

Utilizes four sub-pathways:

- glycolysis
- synthesis of acetyl CoA
- Krebs cycle
- electron transport chain

End result is complete breakdown of glucose to carbon dioxide, water and **ATP**.

Glycolysis

- Occurs in the **cytoplasm**
- Involves splitting of a 6-carbon glucose into two 3-carbon molecules of **pyruvate**
- Great amount of energy remains in bonds of acetyl-CoA
- The Krebs cycle transfers much of this energy to electron carriers NAD⁺ and FAD.
- The Krebs cycle occurs in the matrix of **mitochondria**.

Most of the ATP made in cellular respiration comes from the stepwise release of energy through a series of reactions between molecules known as the electron transport chain (ETC).

The ETC is located in the mitochondria.

How Do We Metabolize Things That Aren't Glucose?

- Fats, proteins, and carbohydrates can all provide energy for the cell.
- The basic pathways used to extract energy from fat and protein are the same as for carbs: glycolysis, Krebs, ETC—but there are some extra steps.

Excess amino acids can be used to synthesize pyruvate, acetyl CoA, and α-ketoglutarate, which enters the Krebs cycle.

- Before fats/triglycerides can be broken down to release energy, they must be converted to smaller units.
- The first step is to break the bonds between the glycerol and the fatty acids.
- Glycerol is a 3-carbon molecule that is converted into **glyceraldehyde-3-phosphate**.
- Because glyceraldehyde-3-phosphate is **involved in one of the steps in glycolysis**, it can enter the glycolysis pathway.
- The long fatty acid molecules (typically 14 to 20 carbons long) must also be processed before they can be further metabolized.
- They enter the mitochondrion, where each long chain of carbons that makes up the carbon skeleton is hydrolyzed (split by the addition of a water molecule) into 2-carbon fragments.
- Each of the 2-carbon fragments is converted into acetyl Coenzyme A, which can then be sent through the Krebs cycle.

Metabolism: Diet and Nutrition

In order to survive, we need the following:

Macronutrients:

- **Carbohydrates**
- **Proteins**
- **Lipids**

Micronutrients:

- **Vitamins**
- **Minerals**
- **Water**

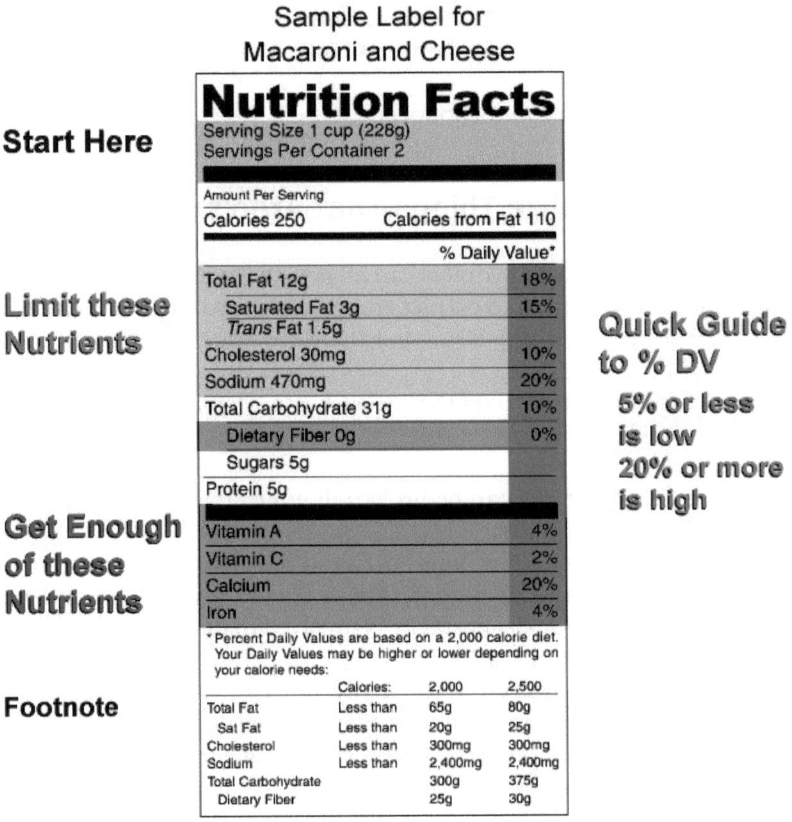

FIGURE 4.9. Nutrition Label

What Is a Calorie?

- **A measure of the energy stored in nutrients**
- **Carbohydrates** have 4 calories per gram.
- **Fiber**, a type of less-digestible carb, has 0 calories per gram.
- **Proteins** have 4 calories per gram.
- **Fats** have 9 calories per gram.

Three of the four macronutrients commonly occur on nutrition labels, and are the major components of our diet.

Carbohydrates

Monosaccharides

- **single** sugars (one molecule)
- simplest
- glucose, fructose, galactose

Disaccharides

- **double** sugars
- combination of two monosaccharides

    - **sucrose** = glucose + fructose
    - **lactose** = glucose + galactose

Polysaccharides

- macromolecules; **polymers** composed of several sugars
- can be the same monomer (many of same monosaccharide) or mixture of monomers

    - **food storage** carbohydrates: **glycogen** (animals), **starch** (plants)
    - **structural** carbohydrates: **chitin** (animals), **cellulose** (plants)

Proteins

Proteins perform many functions in cells.

- **Structural**

    - Components in cell walls, membranes, and within cells themselves

- **Enzymes**

    - Chemicals that speed up a chemical reaction

- **Regulation**

    - Some regulate cell function by stimulating or hindering either the action of other proteins or the expression of genes.

- **Transportation**

    - Some act as channels and "pumps" that move substances into or out of cells.

- **Defense**

    - Antibodies are proteins that defend your body against microorganisms.

Dietary Lipids

Saturated Fats

- Mostly from animal sources

- Single bonds between the carbons in their fatty acid tails (all carbons are bonded to max number of hydrogens possible)
- Hydrocarbon chains fairly straight and packed closely together … so are solid at room temperature.

Unsaturated Fats (Oils)

- Mostly from plant sources
- Have double bonds between some carbons in the hydrocarbon tail, causing bends or "kinks" in shape
- Kinks in the tails prevent unsaturated fats from packing closely together … so they are liquid at room temperature.
- There are two families of essential fatty acids (EFAs): omega-3 and omega-6. Fats from each of these families are essential, as the body can convert one omega-3 to another omega-3, for example, but cannot create an omega-3 from scratch.

Micronutrients

Our diet must also include **essential nutrients** that our bodies cannot manufacture but are needed for biological function, such as the following:

- Vitamins
  - Example: B vitamins
  - Eight water-soluble vitamins that play important roles in cell metabolism

- Minerals
  - Example: Magnesium
  - Magnesium plays an important role in the **production** and **transport of ENERGY**.
  - It is also important for the **contraction** and **relaxation of muscles**, the **synthesis of protein, and assisting certain enzymes** in the body.

The Role of the Cytoplasm and Mitochondria in Cellular Metabolism and Fermentation: Recap

Mitochondria

Mitochondria contain their own DNA (termed mtDNA) passed down from the mother. They function as the sites of energy/ATP production. They have been called the powerhouse of the cell.

Glycolysis (*Lysing* Glucose)

Many reactions make up the process called glycolysis. ALL organisms accomplish glycolysis in their cytoplasm. To begin the process, ATP is "invested" into the reaction. Also, NAD⁺ is converted into NADH + H⁺. The process of glycolysis starts with glucose, a 6-C molecule, and splits it into two 3-C compounds. The end of glycolysis produces two pyruvate (3-C) molecules, with a net gain of 2 ATPs and two NADHs per glucose.

Anaerobic Pathways—Fermentation

Under anaerobic conditions (absence of oxygen) pyruvate can be sent into one of three pathways: lactic acid fermentation (humans ferment pyruvate into lactic acid, mostly in muscles where oxygen becomes depleted during exercise—this lactic acid causes muscle stiffness and pain), alcohol fermentation (yeast), or cellular (aerobic) respiration.

Aerobic Respiration

When oxygen is present (aerobic conditions), humans will send the pyruvate through the Krebs cycle (aka citric acid cycle, which occurs in the matrix of the mitochondria), and then electron transport, to produce a lot of ATP. Pyruvate is first converted to acetyl Co-A, the entry molecule to the Krebs cycle. The acetyl Co-A (2-C) then undergoes multiple reactions resulting in the release of carbon dioxide, and the formation of ATP, GTP (converted to ATP), NADH, and $FADH_2$. The electron transport system (ETS) chemicals are found in the mitochondrial membranes. The higher energy molecules NADH and $FADH_2$ are "cashed in" to produce a lot of ATP. Cytochromes are the molecules that pass the electrons along the ETS chain. ATP and H_2O are formed in this process called oxidative phosphorylation.

The Endomembrane System

- Consists of
    - **Vacuoles and vesicles**
    - **Lysosomes**
    - **Ribosomes**
    - **Endoplasmic reticulum (ER)**
    - **Golgi apparatus**
    - **Nucleus**

Figure 4.10 shows the endomembranous system—a system of internal membranes within eukaryotic cells that divides the cell into compartments, or organelles. It is a transport system for

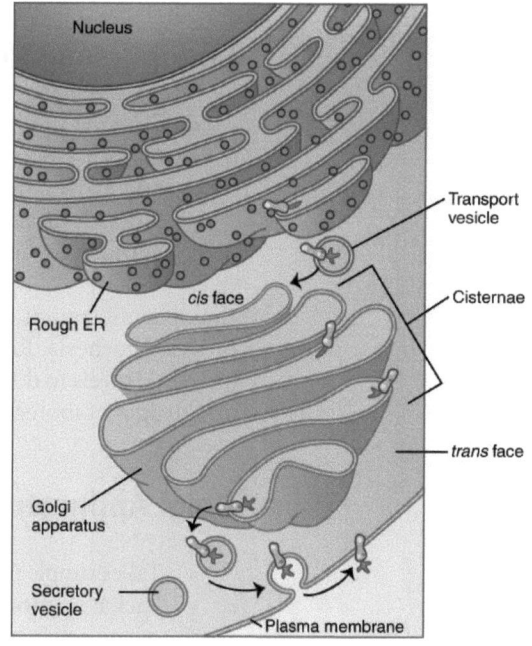

FIGURE 4.10. The Endomembrane System

moving molecules into, out of, and through the interior of the cell, as well as interactive surfaces for lipid and protein synthesis.

Vacuoles and Vesicles

- Store, transport, or digest cellular products and waste
- Small compartments separated from the cytosol by at least one lipid bilayer
- Made in Golgi apparatus, ER, or from parts of the plasma membrane
- Vesicles form while taking in (**endocytosis**) or discharging (**exocytosis**) materials.

Lysosomes

- Formed by the Golgi apparatus
- Break down food into particles and also destroy old cellular components
- Contain hydrolytic enzymes and are involved in intracellular digestion

Ribosomes

Ribosomes are the sites of protein synthesis, in the process of translation. They can be found alone in the cytoplasm, or attached to the endoplasmic reticulum (rough endoplasmic reticulum, or RER).

Endoplasmic Reticulum

- System of membranous channels and vesicles
- **Rough ER** is studded with ribosomes. It is the site of protein synthesis and processing.
- **Smooth ER** lacks ribosomes. It is the site of synthesis of phospholipids and packaging of proteins into vesicles.

Endoplasmic reticulum is made of interconnected membranes that function in protein or lipid synthesis. Rough endoplasmic reticulum (rough ER) is where messenger RNA (mRNA) travels to direct protein synthesis. Smooth ER may be involved in lipid synthesis and drug metabolism, among other possible functions.

Golgi Apparatus

- Takes simple molecules and puts them together into more complex macromolecules
- Packages, modifies, and transports materials to different locations inside/outside of the cell
- Consists of a stack of curved little sacs
- Receives protein and also lipid-filled vesicles from the ER, and packages, processes, and distributes them *within the cell* or for *export out of the cell (secretion)*
- Also encloses digestive enzymes into membranes to form **lysosomes**

Golgi are flattened stacks of membrane-bound sacs which act as a packaging plant, modifying vesicles.

Information Management—The Nucleus

Hereditary material (both DNA and RNA) allows a cell to replicate and/or reproduce. Eukaryotic DNA is organized in structures, called chromosomes, found in the cell nucleus. The nucleus is the location for most of the cell's nucleic acids (DNA and RNA), which control various features of the cell. Deoxyribonucleic acid, DNA, is the physical carrier of inheritance. Ribonucleic acid, RNA, is formed in the nucleus using DNA as a template (transcription). RNA moves out into the cytoplasm where it directs synthesis of proteins by ribosomes (translation). The nucleolus is an area of the nucleus where ribosomes are produced. The nuclear envelope is a double-membrane structure with numerous pores to allow RNA and other chemicals to pass, but the DNA does not leave the nucleus.

Figure 4.11 shows the nucleus of the cell. The nucleus

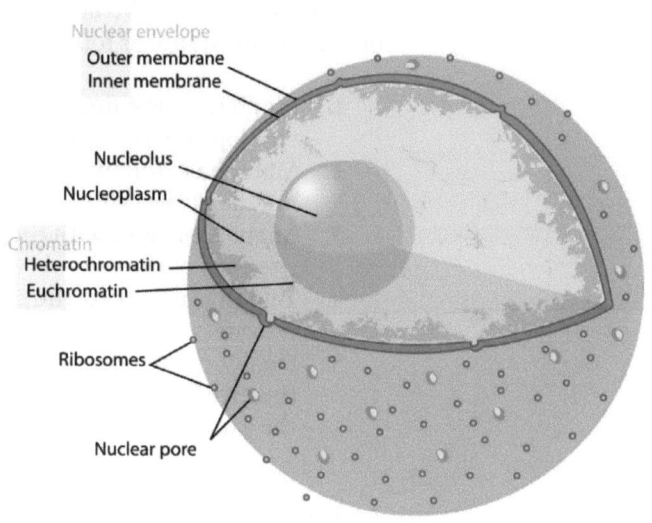

FIGURE 4.11. The Cell Nucleus

- separates the genetic material (DNA) from the rest of the cell, and
- contains DNA, the genetic material, which is a blueprint, or code, for making **proteins.**

The nuclear membrane is the double membrane structure that separates the nucleus from the cytoplasm.

Cell Reproduction

Replication of DNA must occur before a cell reproduces. Cytokinesis (cell splitting) then ends the cell division process. The cell cycle is the sequence of growth1 (G1), DNA replication, growth2 (G2), and cell division that all cells go through. Most cells spend the majority of their time in the stage known as interphase (between phases), the longest part of the cell cycle. After gaining size and ATP, the cells then undergo DNA synthesis (replication of the original DNA molecules). Regulation of the cell cycle must be strictly controlled, or damaged/mutated cells will be produced.

Cancer cells are cells that undergo rapid divisions, producing daughter cells that divide before they have reached functional maturity (and so may contain mutations—mistakes that were not corrected in time). Environmental factors such as temperature, pH, and declining nutrient levels can lead to declining cell division rates.

Mitosis

Mitosis is the process of forming (cloning) identical daughter cells by replicating and dividing the original chromosomes. Replicated chromosomes consist of two molecules of DNA. During mitosis, replicated chromosomes are positioned near the middle of the cytoplasm and then segregated so that each daughter cell receives a copy of the original DNA (if you start with 46 in the parent cell, you should end up with 46 chromosomes in each daughter cell). To do this, cells "pull" chromosomes into each eventual cell. The phases of mitosis are as follows:

1. Prophase is the first stage of mitosis, where chromatin condenses, the nuclear envelope dissolves, and a spindle forms.
2. Metaphase follows prophase and is where the chromosomes migrate to the middle of the cell.
3. Anaphase begins with the separation and the pulling of chromosomes to opposite ends of the cell.
4. Telophase is when the chromosomes reach the ends, the nuclear envelope reforms, chromosomes uncoil into chromatin, and the nucleolus reforms.
5. Finally, cytokinesis is the process of splitting the daughter cells apart (mitosis is the division of the nucleus). There are now two smaller cells, each with exactly the same genetic information as the one original cell.

Meiosis

Haploid and diploid are terms used to refer to the number of sets of chromosomes in a cell. Diploid organisms are those with two (di) sets of chromosomes. Human cells (except for the gametes) are diploid. Haploid cells have only one set of chromosomes. Humans (normally) receive one set of homologous chromosomes (a pair of chromosomes that consists of one chromosome from each parent) during conception. During the formation of gametes (sex cells), the number of chromosomes must be reduced by half, and returned to the full amount when the two gametes fuse during fertilization.

Meiosis is a special type of nuclear division that separates one copy of each homologous chromosome into each new gamete. Meiosis reduces the number of sets of chromosomes by half, so that when fertilization occurs, the full complement of chromosomes of the parents will be reestablished.

Phases of Meiosis

Two successive nuclear divisions occur, meiosis I (reduction) and meiosis II (division). Meiosis produces 4 haploid cells. Mitosis produces 2 diploid cells. Most of the differences between the two processes occur during meiosis I. The phases of meiosis are prophase I (crossing over may occur at this point—chromatids break and may be reattached to a different chromosome); metaphase I; anaphase I; telophase I; prophase II; metaphase II; anaphase II; telophase II. Cytokinesis separates the cells. Thus, mitosis maintains chromosome number, while meiosis reduces it.

Gametogenesis is the process of forming gametes (haploid, "sex" cells) from diploid cells. Spermatogenesis is the process of forming sperm cells by meiosis in specialized gonads (in males—testes). After meiosis, the cells undergo differentiation to become sperm cells. Oogenesis is the process of forming an ovum (egg) by meiosis in specialized gonads (in females—ovaries). In spermatogenesis, all 4 meiotic cells develop into gametes, whereas in oogenesis only one of the cells becomes the large egg (the other cells are termed the polar bodies).

Cell Division—Mitosis and Meiosis: Summary

Figure 4.12 shows the **two major phases** of the cell cycle:

- **G1, S, G2** (3 stages)—DNA uncondensed
- **Prophase, metaphase, anaphase, telophase** (4 stages + **cytokinesis**)

Mitosis

- Division of **somatic (body)** cells (nonreproductive cells) in eukaryotic organisms
- A single cell divides into two identical daughter cells.
- Daughter cells have same number of chromosomes as the parent cell.

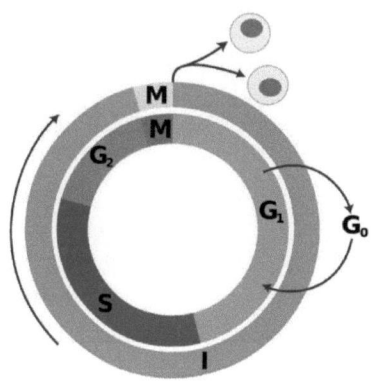

FIGURE 4.12. The Cell Cycle

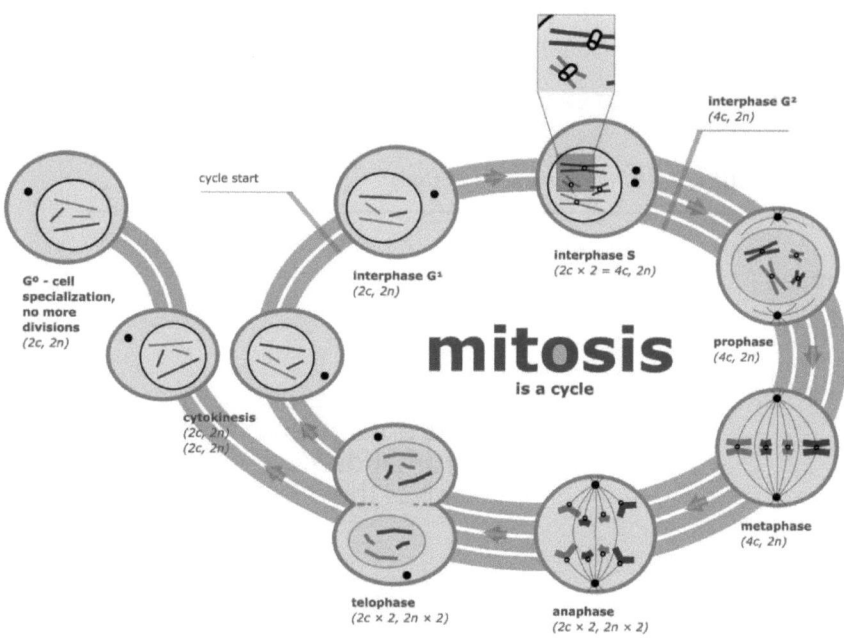

FIGURE 4.13. Mitosis

Figure 4.13 displays the stages of mitosis.

LESSON 4 The Mysteries of Structure | 41

Prophase

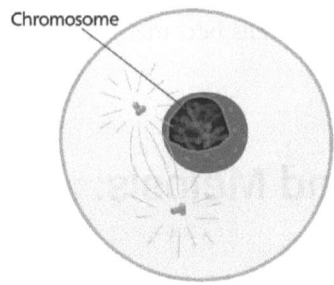

FIGURE 4.14. Prophase of Mitosis

Metaphase

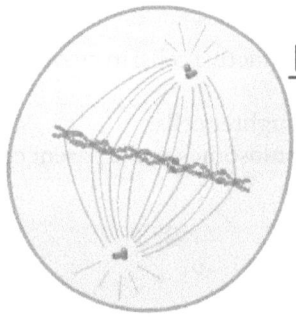

FIGURE 4.15. Metaphase of Mitosis

Anaphase

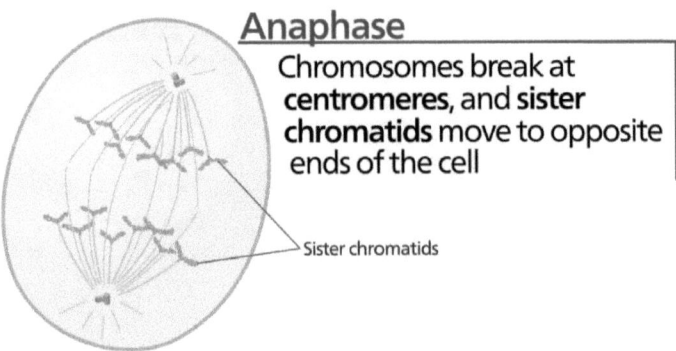

FIGURE 4.16. Anaphase of Mitosis

Telophase

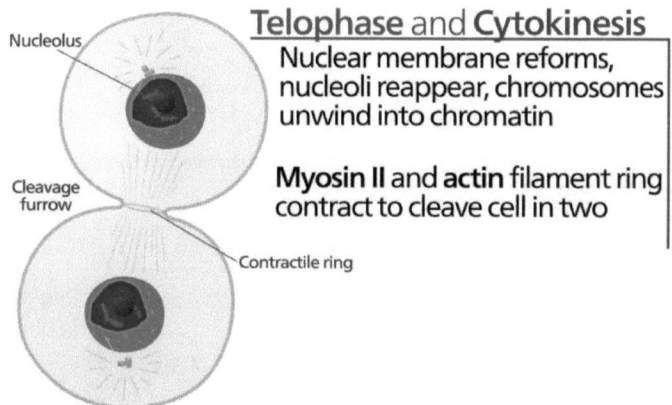

FIGURE 4.17. Telophase and Cytokinesis of Mitosis

Meiosis

Figure 4.18 shows a single germ cell dividing into four unique daughter cells (gametes). Daughter cells have half the number of chromosomes as the parent cell, so they are considered haploid.

- Haploid—one copy of each chromosome, designated as "$\underline{n}$"; the number of chromosomes in one "set"—gametes
- Diploid—two sets of chromosomes; two of each chromosome, designated as "$\underline{2n}$"—somatic cells

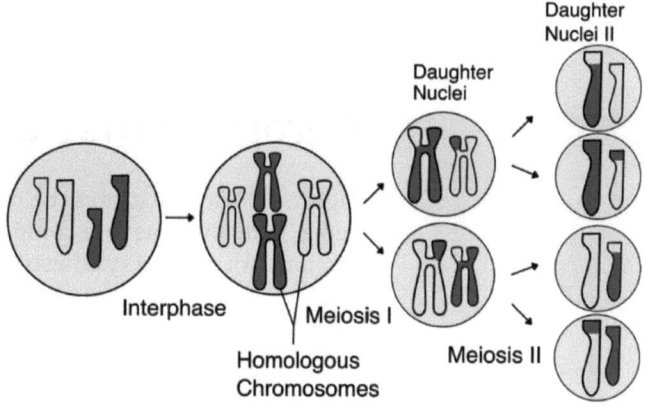

FIGURE 4.18. Meiosis

Sexual Reproduction

- Fusion of two **gametes** to produce a single zygote
- Introduces greater genetic variation, allows genetic recombination
- Zygote has gametes from two different parents.
- At fertilization, 23 chromosomes are donated by each parent (total = 46, or 23 pairs).
- **Gametes** (sperm/ova) contain 22 autosomes and 1 sex chromosome; they are haploid (haploid number n = 23 in humans).

  ○ Fertilization results in diploid zygote.
  ○ Diploid cell, 2n = 46. (n = 23 in humans)

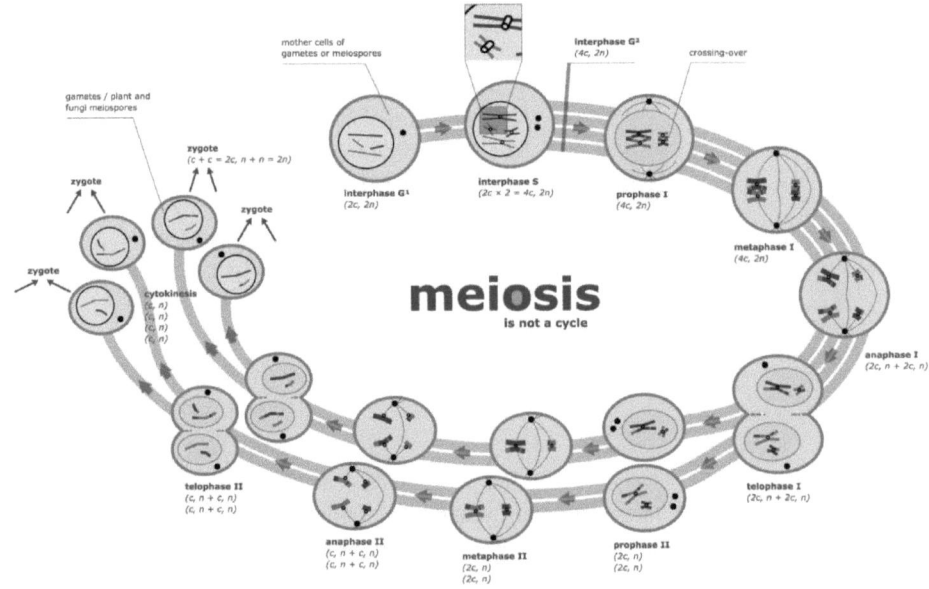

FIGURE 4.19. Meiosis I and Meiosis II

Figure 4.19 shows that there are **two** divisions of the nucleus.

Chromosomes and Genes

Figure 4.20 demonstrates the composition of chromosomes, which are made of many genes, and how genes are composed of introns (noncoding sections) and exons (coding sections, for protein synthesis).

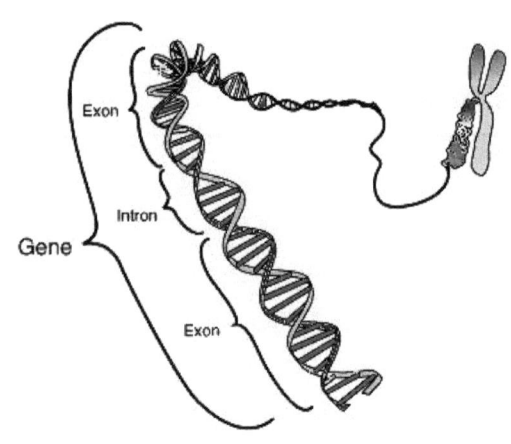

Terminology

- **The genome** is the complete assortment of an organism's DNA.
- Cellular DNA is organized in **chromosomes**.
- **Genes** have specific places on chromosomes.

DNA and Molecular Genetics

DNA is the physical carrier of inheritance. The nucleotide (base, sugar, phosphate) is the basic unit (monomer) of the nucleic acid polymer. There are four nucleotides containing one of the following nucleic acids: cytosine (C), guanine (G), adenine (A), and thymine (T). DNA is a double helix, with the

FIGURE 4.20. Chromosomes and Genes

bases toward the center (like rungs on a ladder) and sugar-phosphate units along the sides (like the sides of a twisted ladder). The strands are complementary—A should always pair with T and C should always pair with G.

Protein Synthesis

One-Gene-One-Protein (This is the old theory, but is it still true?)

The central dogma of molecular biology states that information flows from DNA to RNA via transcription, and then to protein via translation. Transcription is the making of an RNA molecule from a DNA template. Translation is the production of an amino acid sequence (polypeptide) from an RNA molecule (compared to DNA, RNA has ribose sugar instead of deoxyribose sugar; the base uracil (U) replaces thymine (T); and most RNA is single stranded). Messenger RNA (mRNA) is the blueprint for construction of a protein. Ribosomal RNA (rRNA) is the construction site where the protein is made. Transfer RNA (tRNA) delivers the proper amino acid to the proper site at the proper time.

Mutations Redefined

Mutations were defined earlier as any change in the DNA, and any change in DNA is a mistake. A redefinition of a mutation is a change in the DNA base sequence resulting in a change of amino acid(s) in the polypeptide coded for by that gene. Insertion, deletion, or modification of nucleotides can change the polypeptide. Point mutations are the result of the substitution of a single base (e.g., sickle cell anemia). Frame-shift mutations occur when the sequence of the gene is shifted by addition or deletion of one or more bases.

Control of Gene Expression

Each multicellular organism begins as a single-celled organism (zygote), which divides by mitosis. Cells differentiate into functional tissues by turning on some genes while turning off others. The timing of certain gene expressions seems to follow a sequence, such as the production of different types of fetal hemoglobins by mammalian red blood cells, which switch to adult hemoglobin sometime after birth.

The Eukaryotic Genome

The term genome refers to all of the alleles (traits) of an organism. Humans have approximately 3.5×10^9 base pairs (A-T or C-G). Much of the DNA in each cell either has no function or has an unknown function. Humans may use only about 1% of their genome to code for (produce) proteins. Protein-coding sequences are interrupted by noncoding regions. Noncoding regions are known as introns. Coding sequences are known as exons.

Genes, Viruses, and Cancer

Cancer is a disease in which cells no longer control normal cell cycles and growth. Cancer is an inheritable disease (from cells to daughter cells). Once a cell has become cancerous, all of its offspring cells are cancerous. Most carcinogens (cancer-generating factors) are also mutagens (mutation-generating factors). Oncogenes are genes resembling normal genes but in which something has gone wrong, resulting in a cancer. Viruses seem capable of causing cancer.

Molecular Genetics—DNA Replication and Gene Expression: Summary

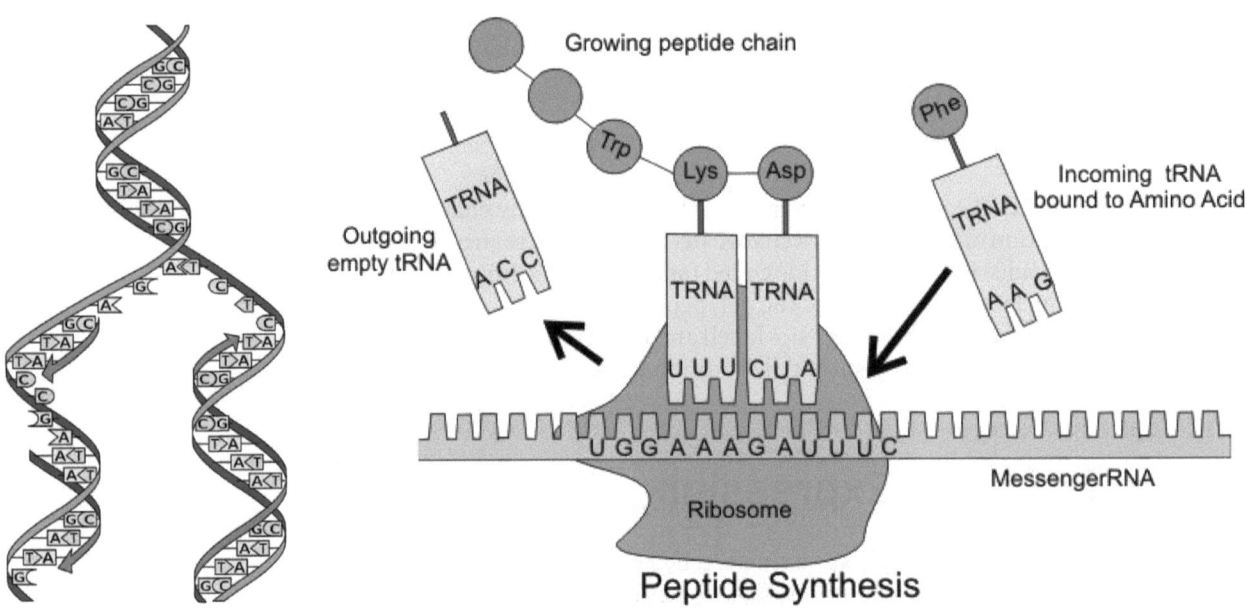

FIGURE 4.21. DNA Replication and Protein Synthesis

Figure 4.21 shows the processes of DNA replication and gene expression through protein synthesis.

DNA Replication—Copying of a Double-Stranded DNA Molecule

- Each DNA strand holds the same genetic information, so each strand can serve as a template for the new, opposite strand.
- Replication occurs prior to cell division, because the new, daughter cell will also need a complete copy of cellular DNA.

Replication Mistakes: Mutations of Genes

- Change in the nucleotide base sequence of a genome; rare.
- Almost always bad news, but …
- Rarely, may lead to a protein having a novel property that improves ability of organism and its descendants to survive and reproduce.

Gene Expression: Transcription and Translation (Making Proteins)

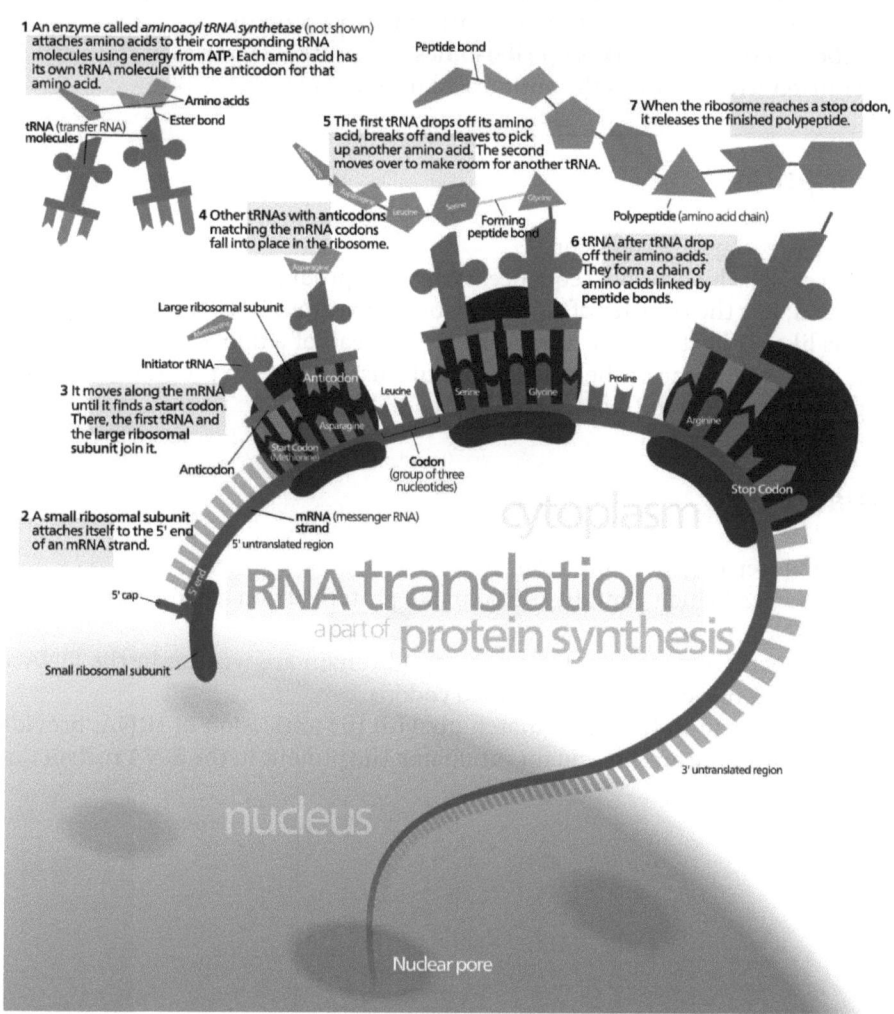

FIGURE 4.22. Transcription and Translation

Figure 4.22 demonstrates the processes of transcription and translation, which allow for the production of proteins based on instruction in a gene coded by DNA. Transcription takes DNA and produces messenger RNA (mRNA), which leaves the nucleus and travels to the ribosome where the mRNA binds to the ribosomal RNA (rRNA) of the ribosome. Transfer RNA (tRNA) then brings in the appropriate amino acid, based on the triplet codon on the mRNA, and starts building the protein (translation).

Genetic Information Copied from DNA Is Transferred to Three Types of RNA

- **Messenger** (mRNA) **is a** copy of information in DNA that is brought to the ribosome, where the information is translated into a protein.
- **Ribosomal** (rRNA) **makes up ribosomes** (the protein factories of the cells).
- **Transfer** (tRNA) brings the amino acid to the ribosome.

Transcription

- First step of gene expression
- Process by which a DNA sequence is copied to produce a complementary mRNA strand; it is the transfer of genetic information from DNA into RNA.
- It is like replication, but produces RNA instead of a duplicate strand of DNA. It is the beginning of the process that ultimately leads to the translation of the genetic code (via mRNA) into a protein.

Translation

- Second step of gene expression
- Ribosomes (which contain rRNA) make proteins from the messages encoded in mRNA.
- The genetic instructions for a polypeptide chain are written in the DNA as a series of 3-nucleotide "words" called codons.
- The codons in mRNA then match up with the anticodon of tRNA, providing the appropriate amino acid for producing the protein in the DNA instructions.

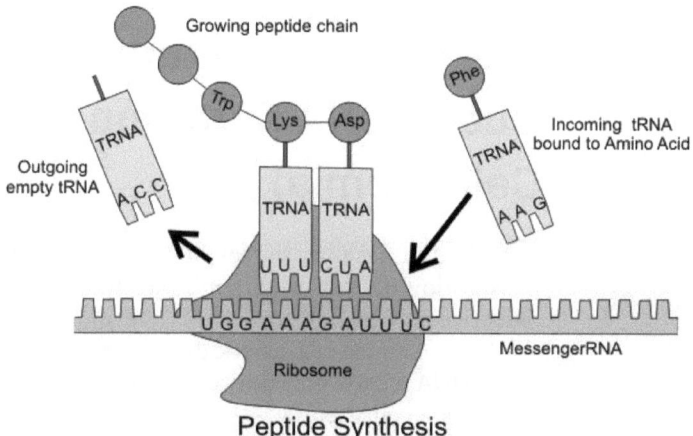

FIGURE 4.23. Translation

Figure 4.23 shows all of the players involved in translation: ribosomes with rRNA, mRNA with the codon based on instruction from DNA, and tRNA with the anticodon allowing the correct amino acid to be brought in during the production of the protein.

Transcription and Translation: Summary

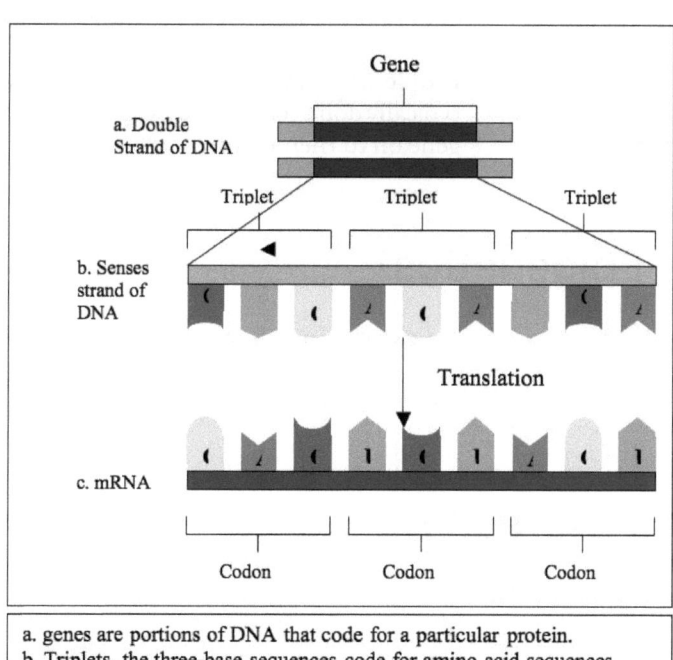

a. genes are portions of DNA that code for a particular protein.
b. Triplets, the three-base sequences code for amino acid sequences.
c. Transcribed mRNA codons, which are complementary to the code in DNA triplets.

FIGURE 4.24. Summary of Transcription and Translation

Example Hereditary Disease Due to DNA Mutation (Sickle Cell Anemia)

Inherited diseases arise from a mistake in the genetic code passed down from a person's parents, thus leading to a defective protein.

As an example of a genetic disorder resulting from an error in the DNA, Figure 4.25 demonstrates the problems associated with sickle cell anemia.

- Most common inherited blood disorder in the United States
- Caused by a mutation in the hemoglobin-beta chain gene
- Hemoglobin transports oxygen from the lungs to other parts of the body.
- Red blood cells with normal hemoglobin (hemoglobin-A) easily flow through blood vessels.
- People with this disease have abnormal hemoglobin molecules that may crystallize and cause red blood cells to become sickle shaped, causing blockages and damaging vital organs and tissue.
- People who are heterozygous for the sickle cell trait typically don't get the disease, but can pass the defective gene on to their children.

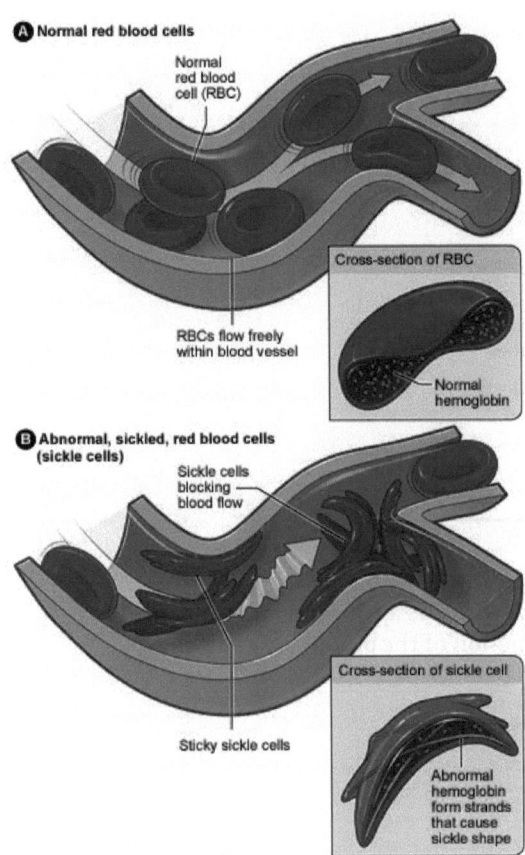

FIGURE 4.25. Sickle Cell Anemia

What Is Epigenetics?

- Heritable changes in **gene expression** that do not involve changes to the DNA sequence.
- A change in **phenotype** without a change in **genotype**.

Genetics (Heredity)

Mendel and His Peas

An Austrian monk, Gregor Mendel (originally schooled in mathematics and statistics), developed the fundamental principles that would become modern-day genetics. Mendel demonstrated that heritable properties are found in discrete units, inherited independently. These eventually were termed genes (alleles).

Mendel's work showed that

- each parent contributes one factor of each trait shown in offspring,
- the two members of each pair of factors separate from each other during gamete (sex cell) formation,
- the blending theory of inheritance was discounted (red + white does not make pink),
- males and females contribute equally to the traits in their offspring, and
- acquired traits are not inherited.

Genetic Terms

- Gene: a unit of inheritance responsible for one trait or character
- Allele: an alternate form of a gene. Usually there are two alleles for every gene.
- Homozygous: when the two alleles are the same
- Heterozygous: when the two alleles are different (dominant allele is expressed)
- Dominant: a term applied to the trait (allele) that is expressed
- Recessive: a term applied to a trait that is only expressed with homozygous alleles
- Phenotype: the physical expression of the alleles for the trait being studied
- Genotype: the allelic composition of an organism
- Punnett squares: probability diagrams illustrating the possible offspring of a mating

Dominant alleles are those that show up in the next generation. Recessive alleles only appear when the individual is homozygous for that recessive trait. Usually homozygous dominant and heterozygous individuals have identical phenotypes, although their genotypes may be different.

Sex-Linkage of Genes

The occurrence of genes on part of the X chromosome, without a corresponding location on the Y chromosome, is referred to as sex-linkage. Sex-linked recessives (hemophilia, baldness, and colorblindness) occur more commonly in males, since there is no chance of them being heterozygous.

The Modern View of the Gene

While Mendel discussed traits, it is now known that genes are segments of DNA that code for proteins. These proteins are responsible for the expression of the phenotype. But, phenotypes can be affected by their environment (epigenetics).

Genes and Chromosomes

Genes are located on specific regions of a certain chromosome, termed the gene locus (a gene is a specific segment of the DNA on a given chromosome). Chromosome abnormalities include insertion, duplication, and deletion (types of mutations). DNA is information, and information has to be in a certain sequence. Insertion, deletion, or duplication of any part of a gene will change the gene product (protein).

Key Points

Organisms are composed of cells. The cell is the basic unit of life. All living organisms are composed of cells, and new cells arise only from preexisting cells.

The plasma membrane is a phospholipid bilayer with embedded/attached proteins. Embedded proteins may serve as receptors, transport channels, and enzymes.

- The plasma membrane separates the inside of the cell from the outside. It is selectively permeable (controls entrance and exit of molecules).
- Diffusion is the movement of molecules from high concentration to low concentration.
- Osmosis is the diffusion of water across a semipermeable membrane from high concentration to low concentration. Hypotonic solutions cause cells to swell due to the uptake of water. Hypertonic solutions cause cells to shrink. In an isotonic solution, the same concentration of solutes and water on both sides of the membrane allows the cells to maintain their size and shape.
- If a protein carrier helps a molecule across the membrane from higher concentration to lower concentration without using energy, it is called facilitated transport.
- Active transport involves a protein that transports a molecule across the membrane from lower concentration to higher concentration using energy (ATP).
- Other transport mechanisms also exist. Endocytosis occurs when a plasma membrane forms a vesicle around a particle as it enters the cell (phagocytosis and pinocytosis). Exocytosis is essentially the reverse of endocytosis (a vesicle fuses with the plasma membrane as a substance is secreted from the cell).

Other intracellular organelles include the following:

- Cilia and flagella are thin projections of cells involved in movement.
- The Golgi apparatus is a stack of vacuoles that packages, stores, and distributes proteins produced by the RER.
- The Golgi apparatus also produces lysosomes (which contain enzymes to digest unwanted materials inside the cell).
- Ribosomes are the sites of protein synthesis in the cytoplasm.
- The endoplasmic reticulum (ER) forms channels in the cytoplasm. Rough ER (with ribosomes) is where protein synthesis occurs. Smooth ER (without ribosomes) is where phospholipids and other compounds are made.
- The nucleus stores genetic information in the chromosomes. The nucleus is separated from the cytoplasm by a double membrane—the nuclear envelope (continuous with the endoplasmic reticulum).

Metabolism includes all the chemical reactions that occur in a cell (anabolism—building, catabolism—breaking down).

- Metabolism requires pathways carried out by enzymes. These pathways are regulated by the cell depending on its metabolic needs (energy excess or energy deficiency).
- Enzymes speed up the rate of a chemical reaction. Enzymes may have nonprotein helper molecules, called coenzymes (dietary vitamins, usually).

- Mitochondria are the organelles that convert chemical energy of fuels into the ATP. They use oxygen for this aerobic metabolism, whereas the cytoplasm can produce ATP anaerobically (without oxygen).

The breakdown of glucose is used to produce ATP. The breakdown of glucose to carbon dioxide and water uses three possible pathways: glycolysis, the citric acid cycle, and electron transport.

- Glycolysis is anaerobic (needs no oxygen) and occurs in the cytoplasm. During glycolysis, glucose breaks down into two 3-carbon molecules of pyruvate. If oxygen is present, pyruvate enters the mitochondria where the citric acid cycle is found. Otherwise, fermentation occurs.
- The citric acid cycle (Krebs cycle) is a series of enzymatic reactions in the mitochondria. Carbon dioxide is released, and hydrogen atoms are sent to electron transport by NADH.
- Electron transport is embedded in the mitochondria. Oxygen is the final acceptor of the electrons at the end of the chain. The energy released is used to produce ATP.
- Fermentation is anaerobic. Pyruvate is converted to lactic acid through fermentation (or ethanol by yeast), producing little ATP, but NAD^+ must be replaced in order for glycolysis to begin again, so it is an important alternative.

DNA and the proteins that assist in the organizational structure of chromosomes are called chromatin. Humans have 46 chromosomes that occur in 23 pairs. Twenty-two of these pairs are called **autosomes**. One pair of chromosomes is called the **sex chromosomes**. Males have the X and Y sex chromosomes, and females have two X chromosomes.

The cell cycle is an orderly process that results in the division of one cell into two identical cells. It has two parts: interphase and cell division.

- Interphase

    - Most of the cell cycle is spent in interphase, which is divided into three stages: G_1, S, and G_2. The G_1 stage occurs before DNA synthesis. The S stage includes DNA synthesis. The G_2 stage occurs after DNA synthesis.

- Mitosis and cytokinesis

    - Cell division, consisting of mitosis (the division of the nucleus) and cytokinesis (the division of the cytoplasm), increases the number of body cells.

Mitosis is replication of the cell. The nuclei of the two new daughter cells have the same number and kinds of chromosomes as the parent cell.

Mitosis is divided into phases: prophase, metaphase, anaphase, and telophase.

- Prophase

    - Spindle fibers appear, the chromosomes condense, the nuclear envelope fragments, and the nucleolus disappears. Spindle fibers attach to the centromeres of the chromosomes.

- Metaphase
  - Metaphase involves the lining up of chromosomes along the cell equator.
- Anaphase
  - At the start of anaphase, sister chromatids split and then are pulled toward respective poles of the cells. The chromatids are now chromosomes.
- Telophase
  - When chromosomes arrive at each pole, telophase begins. In telophase, there are two daughter nuclei.

Cytokinesis is the division of the cytoplasm and organelles.

Meiosis is reduction division. Because meiosis occurs twice, there are four daughter cells, each with half as many chromosomes as the parent cell, the haploid (n) number.

- At the start of meiosis, the parent cell is diploid (2n), and the chromosomes occur in pairs. The members of a pair are called homologous chromosomes, or homologues.
- During meiosis I, chromatids exchange genetic material in crossing over. Next, the homologous chromosomes of each pair separate so that one chromosome from each pair will be in the daughter cell. This reduces the number of chromosomes by half. There is no replication of DNA.
- During meiosis II, the haploid number of chromosomes per cell is still in duplicated condition. This division separates the sister chromatids. Fertilization restores the diploid number of chromosomes in the zygote.
- In prophase I, the spindle appears, nuclear envelopes disappear, homologous chromosomes pair and synapse to form tetrads, and crossing over occurs. This means that chromatids are no longer identical.
- In metaphase I, the homologous pairs line up along the equator.
- Meiosis is part of spermatogenesis, the production of sperm in males, and oogenesis, the production of eggs in females.
- Meiosis is part of gametogenesis, the production of sperm and egg. Meiosis keeps the chromosome number constant from generation to generation. An easier way to keep the chromosome number constant is to reproduce asexually, as unicellular organisms such as bacteria, protozoans, and yeasts do. Binary fission is a form of asexual reproduction. Genetic recombination is the result of meiosis. It occurs because of crossing over and independent alignment of chromosomes.

The genotype represents the actual genes of an individual. Alleles are alternate forms of a gene at the same gene locus. Alleles can be classified as either dominant or recessive. A heterozygote has one recessive and one dominant allele.

The phenotype refers to the outward expression of the genotype (e.g., red hair). Homozygous dominant and heterozygous genotypes result in a dominant phenotype. Only a homozygous recessive genotype results in a recessive phenotype.

Most sex-linked genes (traits) are carried only on the X chromosomes because the Y chromosome is too small. In X-linked traits, the gene is carried on the X chromosome. The Y chromosome is lacking the sex-linked genes that are on the X chromosome. Since males have only one copy of the X chromosome, they show the phenotype for the allele they possess and are thus much more likely than females to show a recessive trait.

- The allele on the X chromosome is shown as a letter attached to the X chromosome. Color blindness is an example of an X-linked trait.

DNA is the genetic material of life and is found in the chromosomes of cells. As genetic material, it must replicate, store information, and undergo change to provide variability.

- DNA is a double helix, composed of two strands that spiral around each other. Each strand is a series of nucleotides. Nucleotides are composed of a phosphate, a sugar, and a base. DNA has the sugar deoxyribose and four different bases: adenine (A), thymine (T), guanine (G), and cytosine (C). There is complementary base pairing within DNA such that A always pairs with T, and G with C. The sugar-phosphate backbone forms the uprights of the DNA double helix, with the base pairs comprising the rungs of the ladderlike shape.
- DNA replication occurs during chromosome duplication. The new DNA molecules are complete, with each parent strand serving as a template for a new strand. A replication error persists as a mutation, which is a permanent change in the sequence of bases.
- RNA (ribonucleic acid) is a single strand of nucleotides containing the sugar ribose. The base uracil (U) replaces thymine in RNA. RNA is a helper molecule in protein synthesis. RNA is divided into messenger RNA (mRNA), ribosomal RNA (rRNA), and transfer RNA (tRNA).

  - Messenger RNA (mRNA) forms off a DNA template in the nucleus and carries genetic information out to the cytoplasm for protein synthesis.
  - Ribosomal RNA (rRNA) is formed off a DNA template in the nucleolus. It joins with proteins imported from the cytoplasm to form the large and small subunits of ribosomes.
  - Transfer RNA (tRNA) transfers amino acids to the ribosomes, where protein is synthesized. There are 20 different types of amino acids used to make proteins.

The first step in gene expression is transcription, when the DNA template is copied into a strand of mRNA. The information from the mRNA is converted into a protein in a process called translation.

- During transcription, a segment of the DNA serves as a template for the production of an RNA molecule.
- During translation, the sequence of codons results in a sequence of amino acids in a protein.

Figure Credits

Fig. 4.1: Source: https://en.wikipedia.org/wiki/File:Animal_cell_structure_en.svg.

Fig. 4.2: Copyright © by Dhatfield (CC BY-SA 3.0) at https://commons.wikimedia.org/wiki/File:Cell_membrane_detailed_diagram_3.svg.

Fig. 4.3: Copyright © by Rlawson (CC BY-SA 3.0) at https://commons.wikimedia.org/wiki/File:Osmosis_experiment.JPG.

Fig. 4.4: Source: https://commons.wikimedia.org/wiki/File:Scheme_facilitated_diffusion_in_cell_membrane-en.svg.

Fig. 4.5: Source: https://commons.wikimedia.org/wiki/File:Scheme_sodium-potassium_pump-gl.svg.

Fig. 4.6: Source: https://commons.wikimedia.org/wiki/File:Animal_mitochondrion_diagram_en.svg.

Fig. 4.7: Source: www.scienceprofonline.com.

Fig. 4.8: Copyright © by RegisFrey (CC BY-SA 3.0) at https://commons.wikimedia.org/wiki/File:CellRespiration.svg.

Fig. 4.9: Source: https://commons.wikimedia.org/wiki/File:Nutrition_label.gif.

Fig. 4.10: Copyright © by OpenStax (CC BY 3.0) at https://commons.wikimedia.org/wiki/File:0314_Golgi_Apparatus_a_en.png.

Fig. 4.11: Source: https://commons.wikimedia.org/wiki/File:Diagram_human_cell_nucleus.svg.

Fig. 4.12: Copyright © by Histidine (CC BY-SA 3.0) at https://commons.wikimedia.org/wiki/File:Cell_Cycle_3-2.svg.

Fig. 4.13: Copyright © by Marek Kultys (CC BY-SA 3.0) at https://commons.wikimedia.org/wiki/File:Mitosis_diagram.jpg.

Fig. 4.14: Copyright © by Kelvinsong (CC BY 3.0) at https://commons.wikimedia.org/wiki/File:Prophase.svg.

Fig. 4.15: Copyright © by Kelvinsong (CC BY 3.0) at https://commons.wikimedia.org/wiki/File:Metaphase.svg.

Fig. 4.16: Copyright © by Kelvinsong (CC BY 3.0) at https://commons.wikimedia.org/wiki/File:Anaphase.svg.

Fig. 4.17: Copyright © by Kelvinsong (CC BY 3.0) at https://commons.wikimedia.org/wiki/File:Telophase.svg.

Fig. 4.18: Copyright © by Rdbickel (CC BY-SA 4.0) at https://commons.wikimedia.org/wiki/File:Meiosis_Overview_new.svg.

Fig. 4.19: Copyright © by Marek Kultys (CC BY-SA 3.0) at https://commons.wikimedia.org/wiki/File:Meiosis_diagram.jpg.

Fig. 4.20: Source: https://commons.wikimedia.org/wiki/File:Gene.png.

Fig. 4.21a: Source: https://commons.wikimedia.org/wiki/File:DNA_replication_split.svg.

Fig. 4.21b: Copyright © by Boumphreyfr (CC BY-SA 3.0) at https://commons.wikimedia.org/wiki/File:Peptide_syn.png.

Fig. 4.22: Copyright © by Kelvinsong (CC BY 3.0) at https://commons.wikimedia.org/wiki/File:Protein_synthesis.svg.

Fig. 4.23: Copyright © by Boumphreyfr (CC BY-SA 3.0) at https://commons.wikimedia.org/wiki/File:Peptide_syn.png.

Fig. 4.24: Source: https://commons.wikimedia.org/wiki/File:Transcription.png.

Fig. 4.25: Source: https://commons.wikimedia.org/wiki/File:Sickle_cell_01.jpg.

Function of Skeleton and Joints

LESSON 5

Objectives

- **State** the functions of the skeletal system.
- **Identify** the bones of the skull, hyoid, vertebral column, and rib cage.
- **Identify** the regions of the vertebral column.
- **Identify** the bones of the pelvic and pectoral girdles.
- **Identify** the bones of the upper and lower limbs.
- **Describe** the structure and operation of a synovial joint.

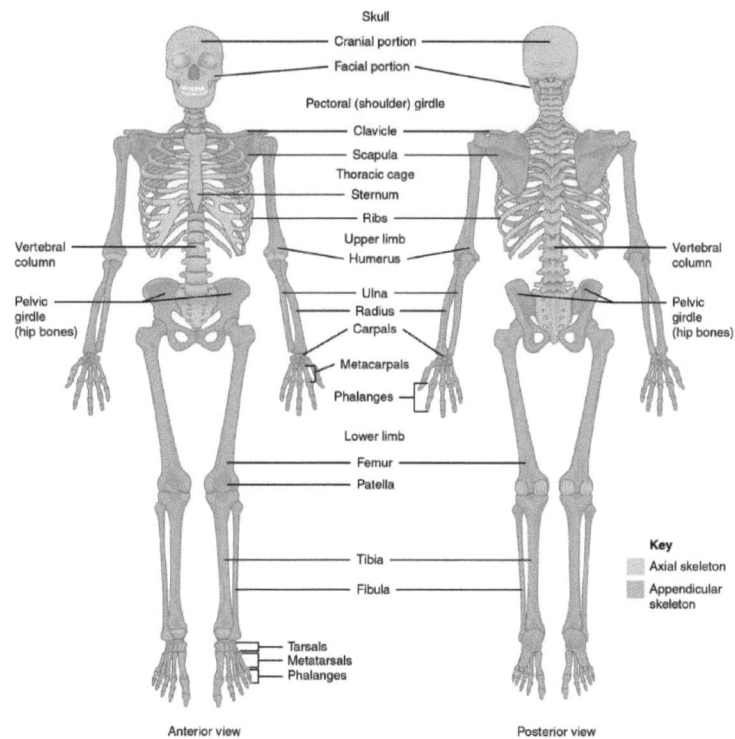

FIGURE 5.1. The Human Skeleton

Figure 5.1 is the human skeleton, highlighting all of the bones of both the axial skeleton and the appendicular skeleton. Not all 206 bones are listed.

- The human skeleton provides the structure of the body.
- Composed of 270 bones at birth, it decreases to 206 bones by adulthood due to some bones fusing together.
- The human skeleton serves six major functions: support, movement, protection, production of blood cells (red bone marrow), storage of ions, and endocrine regulation.

Skeletal System

Humans have developed an internal skeleton composed of bone and cartilage, both a type of connective tissue. Some areas of the human body retain cartilage in the adult (in joints and structures such as the ribs, trachea, nose, and ears). Certain cells in the bones produce immune cells as well as important cellular components of the blood. Bone also helps regulate blood calcium levels.

The Axial and Appendicular Skeletons

The axial skeleton consists of the skull, vertebral column, and rib cage. The appendicular skeleton contains the bones of the appendages (limbs, extremities), and the pectoral and pelvic girdles.

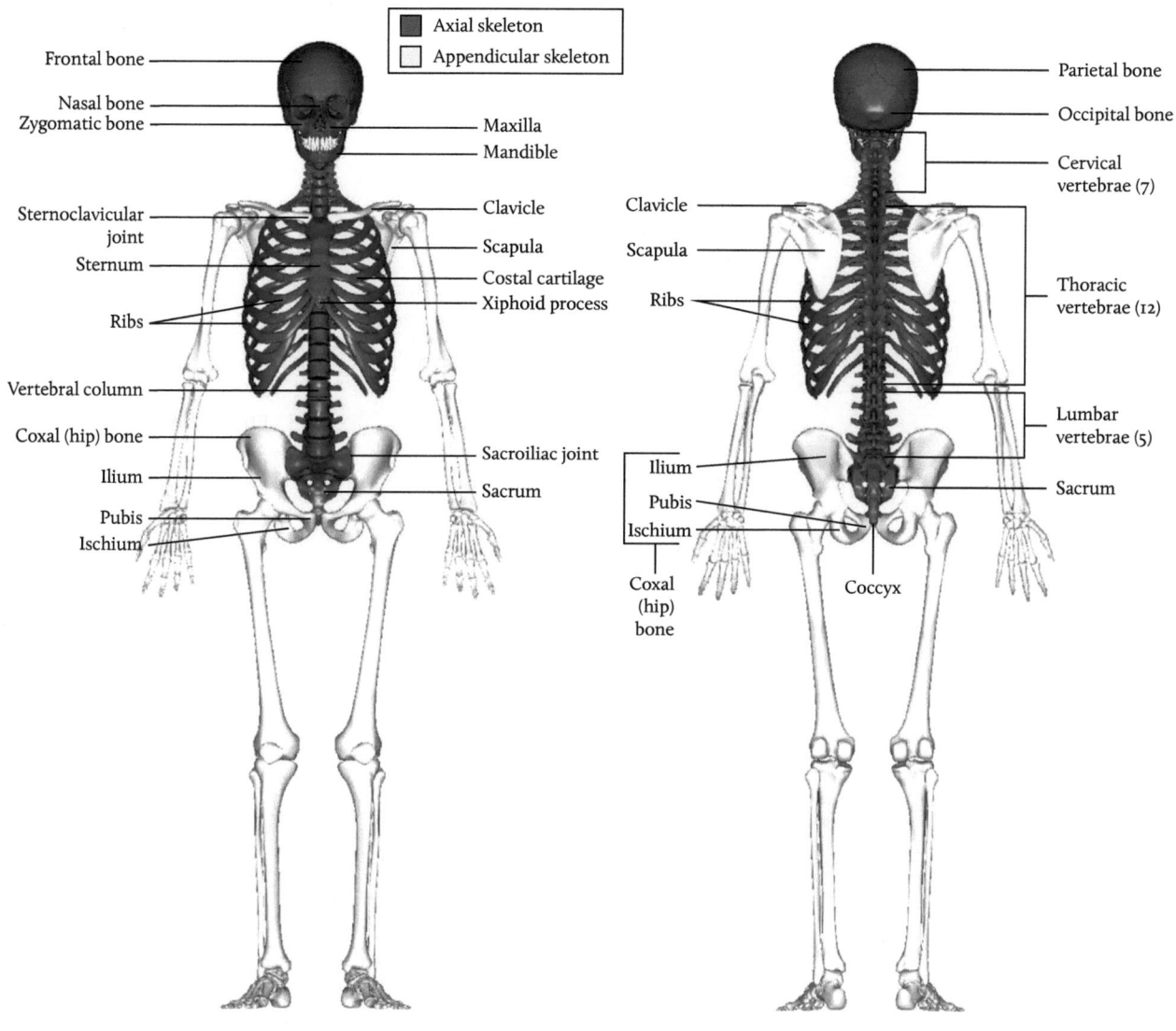

FIGURE 5.2. Axial Skeleton

Axial Skeleton

- The human skull, or cranium, has a few individual bones tightly fitted together at immovable joints. At birth many of these joints are not completely fused together, leading to several "soft spots" or fontanels, which do not completely fuse until about 14 to 18 months.

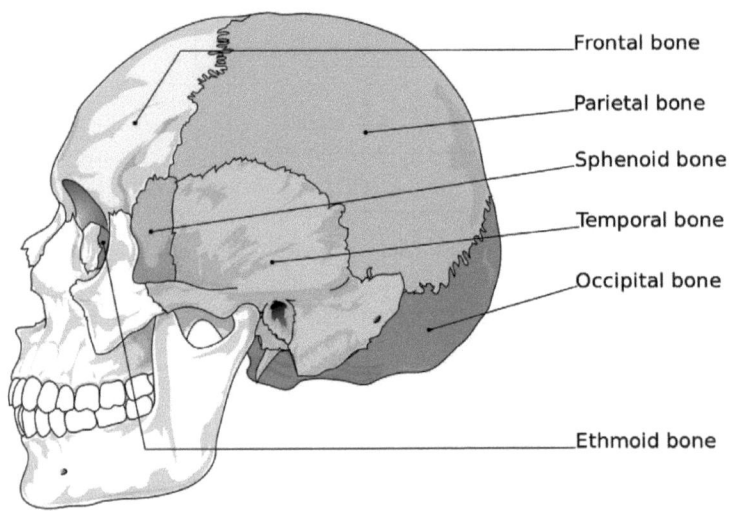

FIGURE 5.3. Bones of the Skull

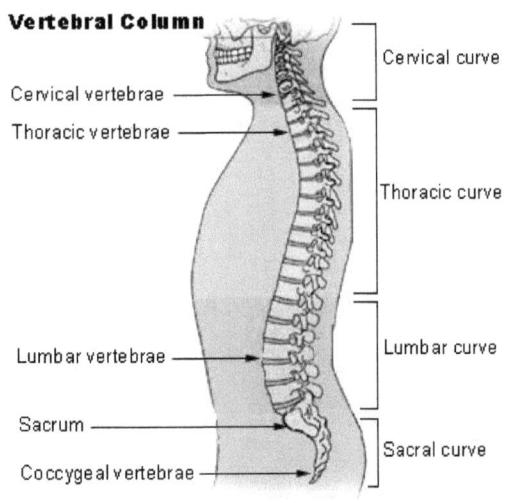

FIGURE 5.4. The Vertebral Column

- The vertebral column has 33 individual vertebrae separated from each other by a cartilaginous disk. These disks allow for some flexibility of the spinal column, although the disks may deteriorate with age. The sternum is connected to all the ribs except the lowest two. Cartilage allows for the flexibility of the rib cage during breathing.

Cervical:

- Seven neck vertebrae, C1–C7
- C1, the first cervical vertebra, is named atlas, C2, the second cervical vertebra, is named axis, and C7, the seventh cervical vertebra, can be felt if you run your hands down the back of your neck—that largest bump is the spinous process of C7).

Thoracic:
- Twelve thoracic vertebrae articulate with the twelve pairs of ribs.

Lumbar:

- Largest vertebrae in lower back

Sacral:

- Five vertebrae that usually fuse during development

Coccygeal:

- Tailbones, 3 to 5 very small vertebrae

Articulate: form a joint

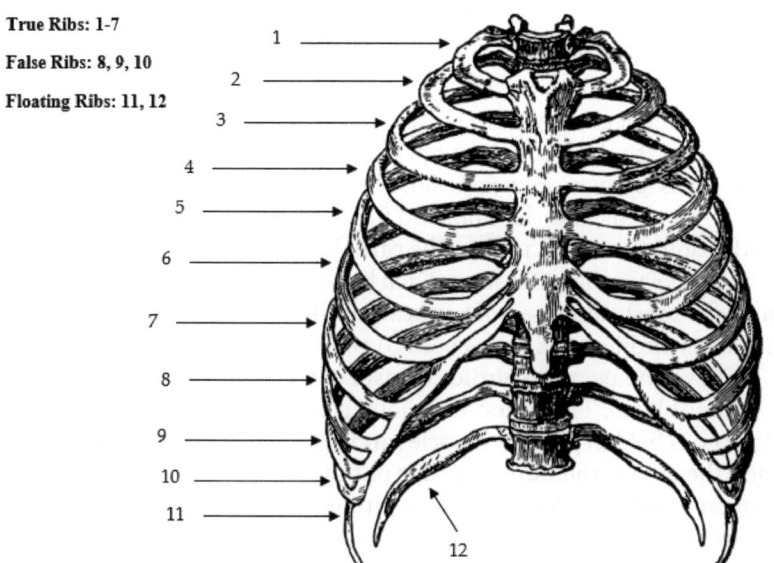

FIGURE 5.5. Ribs

- First 7 ribs are "true" ribs.
- Ribs 8 to 10 are "false" ribs.
- Ribs 11 and 12 are "floating" ribs (not connected to the sternum).

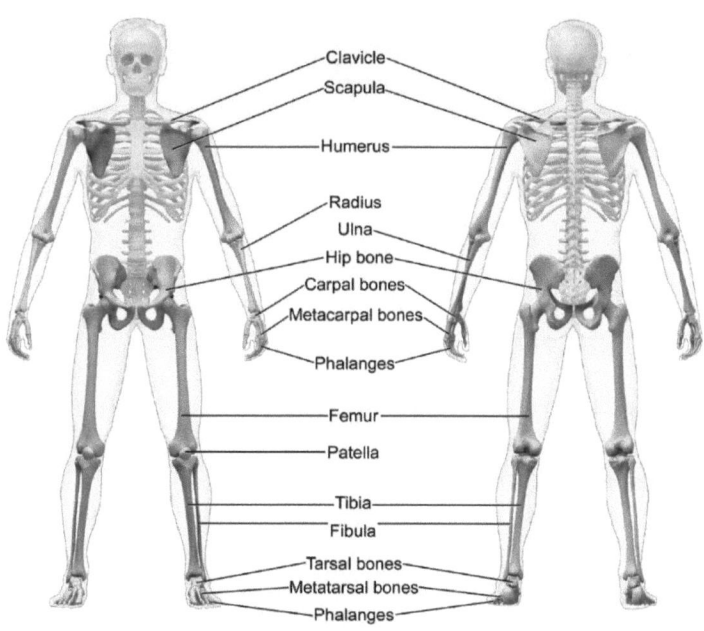

FIGURE 5.6. The Appendicular Skeleton

Appendicular Skeleton

- The arms and legs are part of the appendicular skeleton. The upper bones of the extremities are the humerus (upper arm) and femur (upper leg). Below the arm or leg joints (elbows or knees), those bones articulate (join with) the radius and ulna in the arms and the tibia and fibula in legs; those bones then connect to another joint (wrist or ankle). The carpals make up the wrist joint, whereas the tarsals make up the ankle joint. Each hand or foot ends in five digits (phalanges, fingers or toes) composed of metacarpals (hands) or metatarsals (feet).
- The arms and legs are connected to the rest of the skeleton by bones known as girdles. The pectoral girdle consists of the clavicle (collarbone) and scapula (shoulder blade). The humerus is joined to the pectoral girdle at the shoulder joint and is held in place by muscles and ligaments. A dislocated shoulder occurs when the end of the humerus slips out of the socket of the scapula, stretching ligaments and muscles. The pelvic girdle consists of two hipbones that form a hollow cavity called the pelvis. The vertebral column attaches to the top of the pelvis, whereas the femur of each leg attaches to the bottom. The pelvic girdle transfers the weight of the body to the legs and feet.

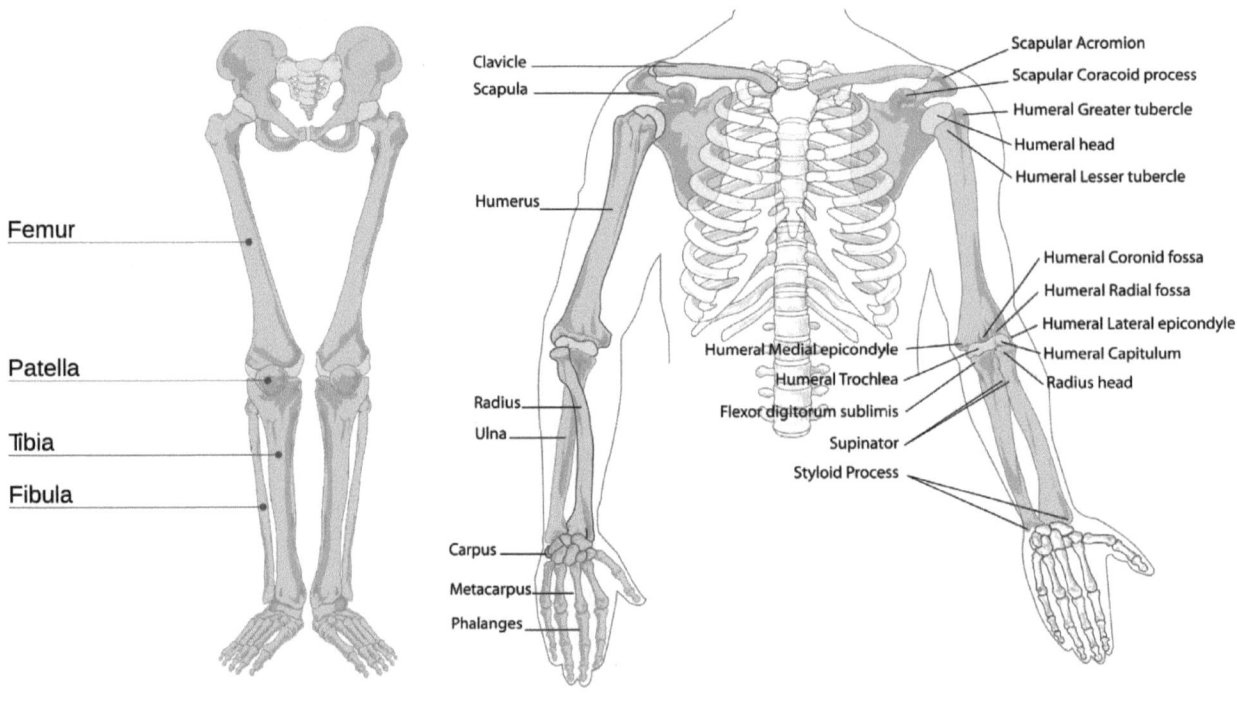

FIGURE 5.7. Appendages

Bone Tissue

- Although bones vary in size and shape, they have certain structural similarities. Bones have cells embedded in calcium and collagen fibers. Compact bone forms the shafts of long bones. Spongy bone forms the inner layer.
- Compact bone has a series of canals surrounded by concentric layers of bone cells (osteocytes) and minerals. New bone is formed by the osteocytes. Spongy bone occurs at the ends of long bones and is less dense than compact bone. The spongy bone of the femur, humerus, and sternum contains red marrow, where stem cells reproduce and form blood and immune cells. Yellow marrow, at the center of these bones, stores fat.

Bone Growth

- Spongy bone forms, and osteoblasts (bone builders) lay down the mineral portions of the bone. Osteoclasts (bone dissolvers) remove material from the center of the bone. Cartilage remains at the ends of the bone (growth plates). During childhood, this cartilage allows for growth of bones. Eventually the elongation of the bones stops and the cartilage is converted into bone.
- Exercise can increase the diameter and strength of bone; inactivity can decrease them. Osteoporosis is a disease that primarily affects older, postmenopausal women.

Joints

- There are three types of joints.

  - Immovable joints, like those connecting the cranial bones, have edges that tightly interlock.
  - Partly movable joints allow some degree of flexibility and usually have cartilage between the bones, such as the vertebrae.
  - Synovial joints permit the greatest degree of flexibility and have the ends of bones covered with a connective tissue filled with synovial fluid, like the hip. The synovial membrane has cells producing synovial fluid that lubricates the joint and prevents the cartilage on the bones from rubbing together. Some joints also have tendons (connective tissue linking muscles to bones). Bursae are small sacs filled with synovial fluid that reduce friction in the joint.

Key Points

The skeletal system consists of bone and cartilage. Ligaments join the bones to one another.

- The skeleton supports the body, protects soft body parts, produces blood cells, stores minerals and fat, and works with muscles to allow body movement.
- The shaft of a bone is called the **diaphysis**. The end of a long bone is called an epiphysis. The epiphyses, at the end of bones, are composed of spongy bone that contains red bone marrow.
- Bone cells are called osteocytes.
- Cartilage is weaker and more flexible than bone, and lacks a direct blood supply. Chondrocytes are cartilage cells.

The axial skeleton lies in the midline of the body. It consists of the skull, hyoid bone, vertebral column, and rib cage.

- The skull is made up of the cranium and the facial bones. The cranium has eight bones, incompletely fused in infants (leaving soft spots, or fontanels). Sinuses are found in the cranium to reduce the weight of the skull and give resonance to the voice. The major bones of the cranium are the frontal bone, two parietal bones, an occipital bone, two temporal bones, a sphenoid bone, and an ethmoid bone.
- The frontal bone of the skull forms the forehead. The lower jaw is made up of the mandible. Zygomatic bones make up the cheekbones, and maxillae form the upper jaw. Two nasal bones form the bridge of the nose.
- The hyoid bone is above the larynx, where it anchors the tongue. The hyoid bone is the only bone in the body that does not connect to (articulate with) another bone.
- The vertebral column supports the head and trunk, protects the spinal cord and nerves, and is a site for muscle attachments. Scoliosis is an abnormal (lateral, or sideways) curvature of the spine. Other curvatures include lordosis and kyphosis.
- Cervical vertebrae are in the neck and include the atlas and axis (C1 and C2). Thoracic vertebrae are in the upper back and attach to the ribs. The lumbar vertebrae are in the lower back. Five sacral vertebrae fuse to form the sacrum. The coccyx (tailbone) is made of four fused vertebrae.
- Intervertebral disks pad the vertebrae. They can herniate and rupture.
- The rib cage is composed of the thoracic vertebrae, the ribs and their cartilages, and the sternum.
- There are 12 pairs of ribs that all connect to the thoracic vertebrae in back. In front, the upper 7 connect to the sternum via cartilage (true ribs) followed by 3 ribs that do not connect directly to the sternum via cartilage (false ribs). The lower two pairs of ribs are called "floating ribs" because they are not attached to the sternum at all.
- The sternum is in the midline of the body and protects the heart and lungs. The bones of the sternum are the manubrium, the body, and the xiphoid process.

The appendicular skeleton consists of the pectoral and pelvic girdles and attached limbs.

- The pectoral girdle is made up of the scapula (shoulder blade) and the clavicle (collarbone). The scapula articulates with the head of the humerus, the upper arm bone. The radius and ulna make up the lower arm. The hand is made up of eight carpal bones, five metacarpal bones, and the phalanges (fingers and thumb).
- The pelvic girdle has two large bones (each of which is made up of the ilium, ischium, and pubic) that are fused at the sacrum and make up the pelvis. The acetabulum is where the femur connects to the pelvis. The thigh contains the

femur, and the lower leg has the tibia and fibula. The femur articulates with the tibia at the knee behind the patella (kneecap). The ankle contains seven tarsal bones, and five metatarsal bones. The toes are called phalanges, like the fingers.
- Joints allow for a variety of movement. Synovial joints have a joint capsule lined with synovial membrane and fluid-filled sacs called bursae to reduce friction.

Bones undergo bone growth and repair. Osteoblasts are bone-forming cells. Osteocytes are mature osteoblasts. Osteoclasts are bone-dissolving cells.

- When the epiphyseal plates fuse, bones can no longer increase in length.
- Vitamin D and growth hormone (GH) affect the growth of the bones. Too little growth hormone in childhood results in dwarfism, whereas too much results in gigantism. Too much GH as an adult results in a condition called acromegaly.
- The actions of osteoclasts and osteoblasts allow the body to regulate the amount of calcium in the blood through dissolution of bones. Parathyroid hormone (PTH) stimulates osteoclasts to dissolve the bone, increasing the amount of calcium in the blood. Calcitonin has the opposite effect, stimulating osteoblasts to remove calcium from the blood to build new bone. Reduced estrogen in postmenopausal women may cause osteoporosis.
- Skeletal bones contain both yellow and red marrow. Fat is stored in yellow marrow. Red bone marrow is where blood cells are produced.

Figure Credits

Fig. 5.1: Copyright © by OpenStax College (CC BY 3.0) at https://commons.wikimedia.org/wiki/File:701_Axial_Skeleton-01.jpg.

Fig. 5.2a: Copyright © by The Database Center for Life Science (CC BY-SA 2.1 JP) at https://commons.wikimedia.org/wiki/File:Axial_skeleton_-_anterior_view.png.

Fig. 5.2b: Copyright © by The Database Center for Life Science (CC BY-SA 2.1 JP) at https://commons.wikimedia.org/wiki/File:Axial_skeleton_-_posterior_view.png.

Fig. 5.3: Source: https://commons.wikimedia.org/wiki/File:Cranial_bones_en.svg.

Fig. 5.4: Source: https://commons.wikimedia.org/wiki/File:Illu_vertebral_column.jpg.

Fig. 5.5: Source: https://www.publicdomainpictures.net/en/view-image.php?image=130313&picture=chest-ribcage.

Fig. 5.6: Copyright © by BruceBlaus (CC BY-SA 3.0) at https://commons.wikimedia.org/wiki/File:Appendicular_Skeleton.png.

Fig. 5.7a: Source: https://commons.wikimedia.org/wiki/File:Human_leg_bones_labeled.svg.

Fig. 5.7b: Source: https://commons.wikimedia.org/wiki/File:Human_arm_bones_diagram.svg.

LESSON 6

The Mysteries of Human-Function Systems

Locomotion—Muscular Functions

Objectives

- **List** the three types of muscles and provide a function for each.
- **Describe** the general structure of a skeletal muscle.
- **Identify** the structures of a muscle fiber.
- **Explain** how the sliding filament model is responsible for muscle contraction.
- **Summarize** how activities within the NMJ control muscle fiber contraction.
- **Summarize** how muscle cells produce ATP for muscle contraction.
- **Distinguish** between fast-twitch and slow-twitch muscle fibers.
- **Summarize** the role of the muscular system in body temperature homeostasis.

Musculoskeletal Systems

We move by the actions of muscles on bones. Tendons attach many skeletal muscles to bones. This allows muscle contractions to move the bones. Muscles generally work in pairs to produce movement: when one muscle contracts, the other relaxes, in a process known as antagonism. Muscles allow the body to act in a coordinated manner. The structure and function of the musculoskeletal system provide for human movement, balance, and growth, thanks to the interaction of the musculoskeletal system with the nervous and endocrine systems. Interestingly, muscles have both electrical and chemical activity. There is an electrical charge across the muscle cell membrane, the outside being more positive than the inside. A stimulus causes a reversal of this charge, causing the muscle to contract.

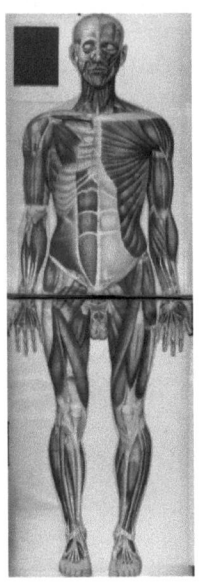

FIGURE 6.1. Muscles of the Human Body

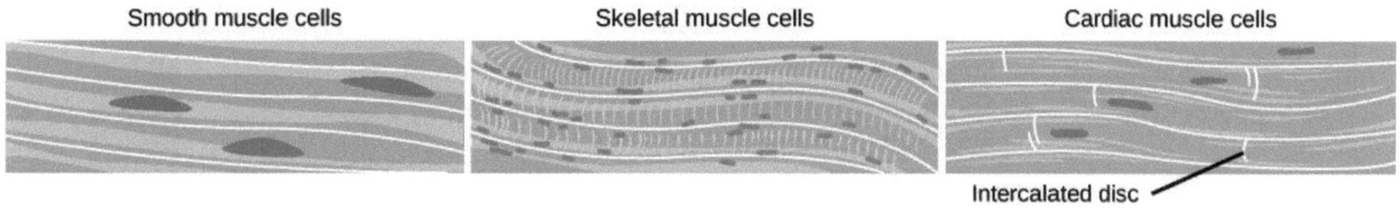

FIGURE 6.2. The Three Types of Muscles

Figure 6.2 show the three types of human muscular tissue: smooth muscle (found in blood vessels and the digestive tract), skeletal muscle (which helps with locomotor activity of the body), and cardiac muscle (involved in contractions of the heart).

Skeletal Muscle Structure

Skeletal muscle fibers have many nuclei. Most of the cell is occupied by striated/striped structures called myofibrils. A sarcomere is the muscle functional unit (a muscle "cell"). Each sarcomere has thick (myosin) and thin (actin) protein filaments. Muscles contract by shortening each sarcomere. The theory of how muscles contract is called the sliding filament model of muscle contraction. Thin actin filaments are on each side of the sarcomere, sliding past each other over the thick myosin filaments until they meet in the middle. Myosin heads attach to binding sites on the actin filaments. The myosin heads detach and then reattach to the nearest active site of the actin filament. Each cycle of attachment, swiveling, and detachment shortens the sarcomere. Hundreds of such cycles occur each second during muscle contraction.

Energy for muscle contraction comes from ATP, the energy "currency" of the cell. ATP binds between the myosin heads and the actin filaments. The release of energy powers the swiveling of the myosin head. Muscles store little ATP and so must recycle the ADP, adenosine diphosphate (ATP minus one high-energy phosphate).

Also, calcium ions are required for each cycle of the myosin–actin interaction. Calcium is released into the sarcomere when a muscle is stimulated by a nerve signal. This calcium uncovers the actin binding sites. When the muscle no longer needs to contract, the calcium ions are pumped out of the sarcomere and back into storage in the sarcoplasmic reticulum (the endoplasmic reticulum of the muscle cell).

Control of Muscle Contraction

Neuromuscular junctions are the point at which a motor neuron attaches to a muscle. Acetylcholine, a chemical transmitter from the motor nerve, is released from the axon terminal when a nerve signal reaches this junction. When the acetylcholine binds to receptors on the muscle cell surface, calcium is released from its storage area in the cell, bringing about a single, short muscle contraction.

Important concepts related to skeletal muscles:

- Structure and function of the muscular system
- Sliding filament theory
- Action of paired muscles

Key Points

The muscular system is involved with movement through muscular contraction.

- Smooth muscle fibers are spindle-shaped cells with a single nucleus. There are no striations. And, contraction is involuntary. Smooth muscle is located in many organs and blood vessels.
- Cardiac muscle is found in the heart. Its fibers also have one nucleus, and are striated. Structures called intercalated disks connect cardiac muscle cells.
- Skeletal muscle fibers are multinucleated, striated, and voluntary.

Skeletal muscles are attached to the skeleton, and their contraction causes the movement of bones at a joint.

- Muscles are often looked at as agonists, synergists, and antagonists. The agonist is the prime mover. Synergists help a muscle perform its activity. Antagonists work against a given muscle's activity.
- Alternating dark and light bands give skeletal muscle a striated appearance.
- The plasma membrane of a muscle fiber is called the sarcolemma. Its endoplasmic reticulum is called the sarcoplasmic reticulum (stores calcium). The sarcolemma invaginates into what are called T tubules that are in contact with the sarcoplasmic reticulum. Calcium ions (Ca^{2+}) are needed for muscle contraction. Contractile units are called sarcomeres. Within sarcomeres, thick filaments are made up of myosin, and thin filaments are made up of actin.
- When a nervous signal reaches a muscle fiber, the sarcoplasmic reticulum releases stored calcium, and the fiber contracts. The myosin filaments pull on the actin filaments, causing them to slide past each other. This is the sliding filament theory.
- The site where a motor neuron contacts a muscle fiber is called the neuromuscular junction (NMJ). When a nerve impulse travels down the neuron, a neurotransmitter called acetylcholine (ACh) is released. The ACh binds to its receptors. The sarcolemma sends signals to the T tubules and sarcoplasmic reticulum. Stored calcium is released, triggering contraction. Special molecules called tropomyosin and troponin combine with the released calcium, removing a "blockade" between the actin and myosin, thus exposing the myosin binding sites, which can now bind to the actin.
- A nerve fiber, together with all of the muscle fibers it innervates, is called a motor unit.
- When some of the muscle's motor units are constantly contracted but not moving, the muscle is firm and solid. It has good tone.

- A muscle has four possible sources of energy. Two of these (glycogen and fat) are stored in muscle, whereas the other two (glucose and plasma fatty acids) are taken from blood.
- Muscle cells store very little ATP. The creatine-phosphate (CP) pathway provides immediate ATP for up to about 10 seconds.
- Anaerobic fermentation produces two ATP molecules from the breakdown of glucose to lactate to provide ATP for about two minutes. The formation of lactate produces muscle pain (burning) and fatigue upon exercising.
- Aerobic metabolism is the slowest but most efficient way to produce ATP. It can use glucose from the breakdown of stored glycogen, glucose from the blood, and fatty acids.
- Fast-twitch fibers (few mitochondria) rely on CP and fermentation (anaerobic, short bursts of powerful muscle contractions; e.g., sprinting, weight lifting), whereas slow-twitch fibers (many mitochondria, and myoglobin) tend to prefer aerobic metabolism of fuels for sustained muscle contractions (e.g., long-distance running, biking, swimming).
- In terms of homeostasis, the muscular system helps to maintain body temperature.

Figure Credits

Fig. 6.1: Source: https://www.nlm.nih.gov/exhibition/historicalanatomies/bougle_home.html.

Fig. 6.2: Copyright © by OpenStax (CC BY 4.0) at https://commons.wikimedia.org/wiki/File:Figure_33_02_12abc.jpg.

Imagine the Possible Brain Functions

LESSON 7.1

Objectives

- **Distinguish** between the central nervous system and the peripheral nervous system.
- **Summarize** the activities that generate and propagate an action potential.
- **Explain** the role of neurotransmitters.
- **Distinguish** between the somatic and autonomic divisions of the PNS.
- **Distinguish** between the sympathetic and parasympathetic divisions of the ANS.
- **Identify** the lobes and major areas of the human brain.
- **Identify** the structures of the brain and provide a function for each.
- **Distinguish** between the functions of the primary motor and somatosensory areas.
- **Identify** the structures of the limbic system.
- **Explain** how the limbic system is involved in memory, language, and speech.

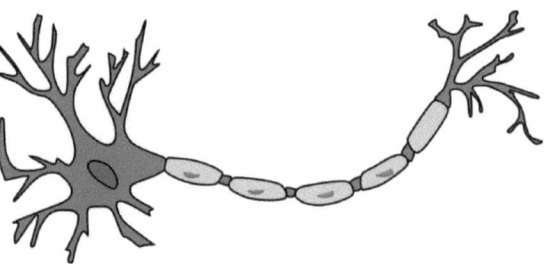

FIGURE 7.1. Neuron

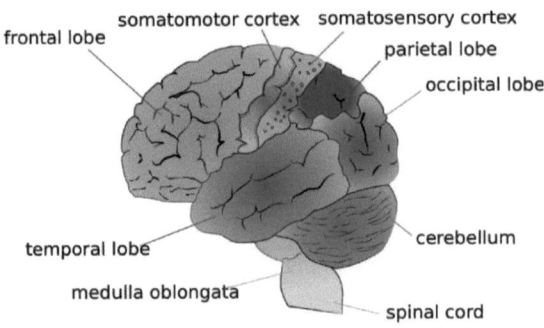

FIGURE 7.2. Brain

Figures 7.1 and 7.2 exemplify the nervous system. The neuron is the signaling cell of the brain and its many areas. The nervous system consists of the brain, spinal cord, sensory organs, and all of the nerves that connect these organs with the rest of the body. Together, these structures are responsible for the control of the body and communication among its parts.

Central Nervous System and the Peripheral Nervous System

The nervous system, the central nervous system (CNS), the peripheral nervous system (PNS), and the autonomic nervous system (ANS) will be explained in terms of composition (CNS = brain and spinal cord, PNS = the rest of the nervous system, ANS = part of the PNS), structure, and the basic functions of nerve cells (both sensory and motor neurons).

The Nervous System

The neuron is the basic unit of the nervous system, and glia are the supporting cells.

Nervous tissue is composed of two main cell types: neurons and glial cells. Neurons transmit the messages. Glial cells often surround neurons. The neuron is the functional unit of the nervous system. You have about a hundred billion neurons in your brain. All neurons have three parts: dendrites, which receive information from another cell and transmit the message to the cell body; the cell body, which contains the nucleus, mitochondria, and other organelles typical of cells; and the axon, which transmits messages away from the cell body.

The types of neurons include sensory neurons, which typically have a long dendrite and short axon, carrying messages from sensory receptors to the central nervous system (referred to as afferent neurons); motor neurons, which have a long axon and short dendrites, and transmit messages from the central nervous system to muscles or glands (referred to as efferent neurons); and interneurons, which connect one neuron to another neuron.

Some axons are wrapped in a myelin sheath (fatty covering) formed from the plasma membranes of specialized glial cells. These glial cells serve to support and nourish the neurons. The gap between these fatty myelin sheaths is known as the node of Ranvier. This allows the electrical signal to jump across and be transmitted faster down the axon than in neurons without myelin sheaths. Signals can jump from node to node hundreds of times faster than signals that travel along the entire surface of the axon.

The Neural Message—Action Potential (An "All-or-None" Phenomenon)

The plasma membrane of neurons has a relatively positive charge on the outside of the membrane compared to the inside of the neuron. This charge difference is called a resting membrane potential (RMP) and is approximately -65 to -70 millivolts. This resting membrane potential results from the difference between sodium and potassium ion concentration across the membrane. Sodium ions are more abundant outside the cell membrane, whereas potassium ions are more abundant inside the cell membrane. This concentration difference is maintained by active transport of ions by the sodium-potassium ATPase pump.

When the charge difference of the membrane is reversed, an action potential (electrical neural signal) occurs. Transmission of the action potential along the membrane occurs until it reaches the axon terminal. The action potential occurs because sodium gates (channels) and potassium gates open sequentially in the cell membrane to allow the ions to diffuse across. Sodium enters first. Then after sodium channels close, potassium channels open and potassium ions leave the inside of the cell. Reversal of ions that crossed the membrane must be set back to the initial resting situation. And that is accomplished with the help of the sodium-potassium ATPase pump.

The action potential begins at one spot on the membrane. It then spreads to adjacent areas of the membrane, transmitting the electrical message along the entire cell membrane. After the action potential passes, there is a refractory period (in which another action potential cannot be generated) at that portion of the membrane. This happens to prevent the message from being transmitted backward along the membrane, making this signal transmission one way.

Generation of an Action Potential Reviewed

- At rest, the outside of the membrane is more positive than the inside.
- Sodium moves through sodium channels into the cell, causing a change in potential; the inside of the axon becomes more positive than the outside (depolarization).
- Potassium ions flow through potassium channels to leave the cell, repolarizing the axon (making the inside more negative than the outside).
- Sodium ions are pumped out of the cell and potassium ions are pumped back into the cell, thus restoring the original resting membrane potential.

Synapses—Electrical to Chemical Transmission of the Action Potential Signal

The gap between one nerve cell and another cell is called a synapse. Messages travel down the axon as an electrical action potential. To get the message across the synapse, chemical neurotransmitters are used. Neurotransmitters are stored in small synaptic vesicles at the tip of the axon (axon terminal).

Arrival of the action potential causes some of these vesicles to move to the end of the axon and release their contents into the synapse. These released neurotransmitters diffuse across the synapse, bind to receptors on the other side, and cause something to happen at the next cell (e.g., ion channels on that cell open). Some neurotransmitters are excitatory and cause a new action potential to be generated; others are inhibitory, preventing generation of a new action potential.

Neurotransmitters cannot be allowed to sit in the synapse indefinitely (think about the consequences, like a muscle contracting indefinitely), so they may be destroyed by synaptic enzymes, they may diffuse out of the synapse, or they may be taken back up by the releasing cell. Acetylcholine is an example of a neurotransmitter destroyed by synaptic enzymes, as is norepinephrine and serotonin, both removed by reuptake into the presynaptic cell. Clostridium botulinum (botulism) produces a toxin found in improperly canned foods and is used for cosmetic purposes (Botox® treatments). This toxin prevents the release of acetylcholine, which can be fatal in certain situations. The bacterium Clostridium tetani (tetanus) produces a toxin that prevents the release of GABA, an important inhibitory neurotransmitter. A number of drugs also operate in the synapses, such as cocaine, LSD, caffeine, and insecticides.

Nervous System Functions

We must constantly be monitoring and maintaining a constant internal environment (homeostasis), as well as monitoring and responding to the external environment. This is accomplished by the three basic functions performed by nervous systems:

- Receive sensory input from the environment
- Integrate/interpret this input
- Respond to the input stimuli with motor output

Sensory Input

Receptors found throughout the body sense changes (including pressure, temperature, pain, body position, taste, sound, light, blood pH, or hormone levels) in the environment. Sensory input comes in many forms and can then be converted to a signal, which is sent to the brain or spinal cord to be interpreted.

Integration and Output

In the sensory centers of the brain, or in the spinal cord, the sensory input is integrated, or interpreted, and leads to a response. The response, typically a motor response, is a signal sent to organs or glands that can convert this signal into a form of action (movement, change in heart rate, release of hormones, etc.).

Divisions of the Nervous System

The central nervous system (CNS) includes the brain and spinal cord. The peripheral nervous system (PNS) is everything else. These nervous systems communicate with almost every organ system in the body through positive and negative feedback loops. Somatic (body) pathways send environmental signals to the brain (afferent). Motor pathways send signals from the brain (efferent) and are either somatic (skeletal) or autonomic (smooth muscle, cardiac muscle, and glands). The autonomic system is subdivided into the sympathetic (SNS) and parasympathetic systems (PSNS).

The Peripheral Nervous System (PNS)

The peripheral nervous system (PNS) contains only nerves and connects the brain and spinal cord (CNS) to the rest of the body. Most of the axons and dendrites of the PNS are surrounded by a white myelin sheath. Groups of cell bodies (nuclei, ganglia) are mostly found in the central nervous system. Cranial nerves in the PNS take impulses to and send signals from the brain (CNS). Spinal nerves take impulses to the spinal cord from peripheral receptors, and send signals away from the spinal cord to effectors (muscles, glands). There are two major subdivisions of the PNS motor pathways: the somatic (body) and the autonomic (automatic; it is further subdivided into the sympathetic and parasympathetic systems). Most sensory input carried in the PNS to the CNS is below a conscious awareness level. Input that does reach the conscious level contributes to our perception of the environment.

The Somatic Nervous System (SNS)

The somatic nervous system includes all nerves controlling the muscular system and many sensory receptors. Sense organs, like the skin, contain receptors, which send information into the CNS. Muscle fibers and various glands are effectors, which receive signals from the CNS. A reflex arc is an automatic, involuntary reaction to a stimulus (e.g., when the doctor taps your knee with a rubber hammer, they are testing your knee-jerk reflex). The reaction to the stimulus is involuntary. The CNS is informed but not in control of the response. Sensory input from the PNS that is not reflexive is processed by the CNS, which then responds with signals to the PNS, which sends these signals to the organs/glands of the body.

The Autonomic Nervous System (ANS)

> Innervate: supply an organ or other body part with nerves

The autonomic nervous system is the part of PNS that controls internal organs. It has two components: sympathetic and parasympathetic. Together they affect the cardiovascular system (heart and vessels) as well as the smooth muscle throughout the gastrointestinal, respiratory, urinary, and reproductive systems. The sympathetic nervous system is involved in the fight, fright, or flight response. The parasympathetic nervous system is involved in the rest-and-digest response. Each of them typically opposes the other (antagonism) at any given organ. Both systems usually innervate the same organs and act in an opposite manner to maintain homeostasis (the sympathetic system may cause your heart to beat faster, whereas the parasympathetic system will slow down the heart).

The Central Nervous System (CNS)

Composed of the brain and spinal cord, the CNS is surrounded by bone (skull and vertebrae) for protection. Cerebrospinal fluid (CSF) and membranes (meninges) also help protect the brain and spinal cord. The brain has three parts: the cerebrum (forebrain), the cerebellum ("baby" cerebrum), and the hindbrain (midbrain and medulla oblongata). The medulla oblongata, part of the hindbrain, is like an extension to the spinal cord.

It is involved with the regulation of involuntary vital activities (heartbeat, breathing, vasoconstriction/blood pressure), and contains reflex centers for vomiting, coughing, sneezing, swallowing, and hiccupping. The midbrain and pons are structures beyond the medulla towards the cerebrum. The midbrain connects the hindbrain and forebrain. The forebrain consists of the diencephalon (thalamus and hypothalamus) and the cerebrum. The hypothalamus, within the cerebrum, regulates homeostasis (thirst, hunger, body temperature, water balance, and blood pressure). It coordinates links between the nervous system and the endocrine system. The thalamus serves as a relay station for most incoming messages. The cerebrum, the largest part of the human brain, is divided into left and right halves (hemispheres) connected to each other by the corpus callosum (a wide collection of axons), which allows for communication between left and right hemispheres. The hemispheres are covered by a thin layer of gray matter known as the cerebral cortex. The cerebrum is responsible for higher-order functions, such as intelligence and reasoning, learning and memory, as well as many other functions. Folds divide the cortex into four lobes: occipital (back), temporal (lower side), parietal (upper side), and frontal (front), none of which really functions alone. Many functions of various parts of the four lobes have been determined, but they do not work without the others. The occipital lobe (back of the head) receives and processes visual information. The temporal lobe (above the ears in the brain) receives auditory signals, processing language and the meaning of words. The parietal lobe (above the temporal lobes on the side of the brain) is associated with the sensory cortex and processes information about touch, taste, pressure, pain, and temperature. The frontal lobe (behind the forehead) is mostly involved in three functions: motor activity and integration of muscle activity, speech, and thought processes. And, finally, the cerebellum is the second largest part of the brain (located below the occipital cortex). Its functions include muscle coordination and maintenance of muscle tone and posture. This allows the cerebellum to coordinate balance. Beyond the brain, the spinal cord runs along the dorsal (back) part of the body and ties the brain to the rest of the body. The gray matter of the spinal cord consists mostly of cell bodies and dendrites. The surrounding white matter is made up of bundles of axons (tracts). Some tracts are ascending (carrying messages to the brain, mostly sensory), whereas other tracts are descending (carrying messages from the brain, mostly motor). The spinal cord is also involved in reflexes that do not necessarily involve the brain.

Specific pathology of certain areas of the brain shows, for example, that most people have their language and speech areas in the left hemisphere. Language comprehension is found in an area of the temporal lobe called Wernicke's area. Speaking ability is in the frontal lobe in an area called Broca's area. Damage to Broca's area causes speech production problems but not problems of language understanding. Lesions in Wernicke's area cause inability to understand written and spoken communication, though the individual still maintains the ability to produce words (though usually incomprehensible). The remaining parts of the cortex are associated with other higher processes, such as planning, memory, personality, thought, emotions, and other human activities.

Key Points

The nervous system is made up of the central nervous system (brain and spinal cord), and the peripheral nervous system (nerves of the body). The nervous system functions to receive sensory input, process and integrate information, and generate motor output.

- Nervous tissue includes neurons that send out signals and neuroglia that support neurons.
- Sensory receptors send information to the central nervous system (CNS) via sensory neurons. Motor neurons send information from the CNS to muscles or glands.
- Neurons have dendrites, a cell body, and an axon, and many of the axons are covered by a fatty myelin sheath. Each of these has small gaps in between called nodes of Ranvier.

Nerve signals send information throughout the nervous system via electrical signals called action potentials.

- An axon's interior is negative (-65 mV to -70 mV) relative to the outside. This is termed the resting membrane potential (RMP), or just resting potential.
- An action potential is a rapid change in the nerve cell's membrane potential. It is an all-or-none phenomenon that occurs when the potential reaches the threshold potential.
- Sodium channels open first and Na^+ flows into the axon, causing depolarization.
- Potassium channels open next, after Na^+ channels have closed, and K^+ flows out of the axon, causing repolarization.
- The action potential occurs in each section of an axon. In myelinated fibers, the action potential jumps between the nodes of Ranvier.
- The sodium-potassium ATPase can actively pump Na^+ out of an axon and pump K^+ into an axon to return the potential back to its resting state.
- Transmission of the nerve signal continues at a synapse (gap) when a neurotransmitter is released.
- The binding of the neurotransmitter to receptors on the postsynaptic cell may cause excitation or inhibition.
- A neurotransmitter released into the synapse must be destroyed by enzymes, diffuse out of the synapse, or be taken back up into the releasing axon terminal.
- Many different neurotransmitters exist. Drugs may work by interfering with the action of these neurotransmitters, or by prolonging neurotransmitter actions.
- Excitatory signals depolarize receiving cells; inhibitory signals hyperpolarize receiving cells.
- Integration of these signals involves summing them up. If the neuron receives more inhibitory signals than excitatory signals, the axon will not reach threshold, and so will not fire an action potential. If a neuron receives more excitatory signals than inhibitory signals, it will transmit a new signal.

The CNS consists of the brain and spinal cord, and receives sensory input and integrates and then sends out motor responses.

- The gray matter of the brain and spinal cord contains cell bodies.
- The white matter of the spinal cord contains myelinated axons. Tracts of axons cross over in the spinal cord, so that the left side of the brain controls the right side of the body and the right side of the brain controls the left side of the body.
- The spinal cord carries out reflex actions, sends sensory information to the brain, and receives motor output from the brain.

- The spinal cord has many reflex arcs. (When blood pressure falls, receptors in the carotid arteries and aorta generate nerve signals, which pass into the spinal cord, then up to the cardiovascular center in the brain, followed by nerve signals passing down the spinal cord, and resulting in the constriction of blood vessels.)
- The brain has four ventricles (two lateral ventricles, the third ventricle, and the fourth ventricle).
- The cerebrum is the largest part of the brain. It receives the sensory input and sends out motor responses after integration occurs.
- The cerebrum has two hemispheres connected by a bundle of nerves called the corpus callosum. Sensation, reasoning, learning and memory, and language/speech take place within the cerebrum. The frontal, parietal, occipital, and temporal lobes each are associated with specific functions.
- The cerebral cortex is the region of the brain that manages sensation, voluntary movement, and all of the thought processes associated with what we call consciousness.
- The primary motor area is in the frontal lobe and sends out motor signals via motor neurons.
- The primary somatosensory area is in the parietal lobe and receives sensory information from sensory receptors.
- Association areas are where integration occurs.
- The prefrontal area of the frontal lobe is concerned with higher mental functions (executive functions).
- Broca's area (speaking and writing) and Wernicke's area (understanding the written and spoken word) are important association areas for language.
- Gray matter deep within the white matter of the cerebral cortex, called basal ganglia, integrates motor commands as secondary motor systems.
- The diencephalon sits by the third ventricle. It contains the hypothalamus (which controls homeostasis) and the thalamus (the central relay station, sending sensory input to the cerebrum). The pineal gland of the diencephalon secretes melatonin to maintain circadian (daily) rhythms.
- The cerebellum receives input from the eyes, ears, joints, and muscles about the position of body parts (proprioception). It also receives motor information from the cerebral cortex about where these parts should be located (close your eyes and touch the tip of your nose with your finger). It, like the basal ganglia, also sends motor signals to skeletal muscles and acts as part of a secondary motor system.
- The brain stem contains the midbrain, the pons, and the medulla oblongata. The medulla oblongata and pons have reflex centers for vital functions, like breathing and heartbeat.
- The limbic system, comprised of several brain structures, includes the hippocampus (involved in the higher mental functions of learning and memory) and the amygdala (involved in emotional processing like fear—PTSD)
- The hippocampus signals the prefrontal area regarding past experiences.
- The amygdala associates experiences with emotions (especially fear).
- Alzheimer disease is a brain disorder characterized by a gradual loss of memory and is intimately associated with the hippocampus.
- Memory allows you to retain a thought or recall an event or other information from the past. Learning takes place when we obtain and utilize knowledge, past memories, and experiences.

- Language depends on semantic memory. Language and speech are dependent upon Broca's area (motor speech area) and Wernicke's area (sensory speech/understanding area). These two areas are in constant communication.

The peripheral nervous system contains only nerves (bundles of axons) and ganglia (cell bodies) involved with cranial nerves and spinal nerves. The 12 cranial nerves are concerned with the head, neck, and facial regions of the body. The vagus nerve (cranial nerve, CN X) travels to most of the internal organs. The 31 pairs of spinal nerves contain both sensory and motor fibers.

- The nerves in the somatic nervous system go to/from the skin, skeletal muscles, and tendons, taking sensory information from sensory receptors to the CNS and sending motor signals from the CNS to skeletal muscles and glands. Automatic responses to a stimulus in the somatic region are called reflexes.

The autonomic nervous system (ANS) controls the activity of cardiac muscles, smooth muscles of organs, and glands.

- The ANS system is divided into the sympathetic and parasympathetic nervous systems.
- The sympathetic division is associated with responses that occur during times of stress (fight, flight, or fright). The neurotransmitter released is primarily norepinephrine (NE).
- The parasympathetic system is associated with responses that occur during times of relaxation (rest and digest). The neurotransmitter is acetylcholine (ACh).

Figure Credits

Fig. 7.1: Source: https://pixabay.com/vectors/neuron-nerve-cell-axon-dendrite-296581/.
Fig. 7.2: Copyright © by Jkwchui (CC BY-SA 3.0) at https://commons.wikimedia.org/wiki/File:Cerebrum_lobes.svg.

Special Senses

LESSON 7.2

Objectives

- **Explain** the role of somatic sensory receptors: touch, pressure, temperature, proprioception, and nociception (pain).
- **Compare** and contrast the senses of taste and smell.
- **Summarize** how the brain receives taste and odor information.
- **Identify** the structures of the human eye.
- **Explain** how the eye focuses on near and far objects.
- **Describe** the role of photoreceptors in vision.
- **Summarize** the abnormalities of the eye that produce vision problems.
- **Identify** the structures of the ear that are involved in hearing.
- **Summarize** how sound waves are converted into nerve signals.
- **Describe** the pathway of sensory information from the ear to the brain.
- **Explain** how mechanoreceptors are involved in the sense of equilibrium.
- **Identify** the structures of the ear involved in the sense of equilibrium.
- **Distinguish** between rotational and gravitational equilibrium.

Sensation and Sensory Receptors

Sensory receptors may be mechanoreceptors (hearing, balance, stretching), photoreceptors (light), chemoreceptors (smell and taste mainly, but also senses connected to the respiratory and circulatory systems), nociceptors (pain) and thermoreceptors (temperature).

Special Senses

Input to the nervous system is in the form of our somatic senses: touch, pressure, pain, temperature, and proprioception (limb position in space), as well as the special senses of vision, taste, smell, and hearing. Sensory input to the brain begins with sensors reacting to stimuli, which are then converted to action potentials sent to the CNS.

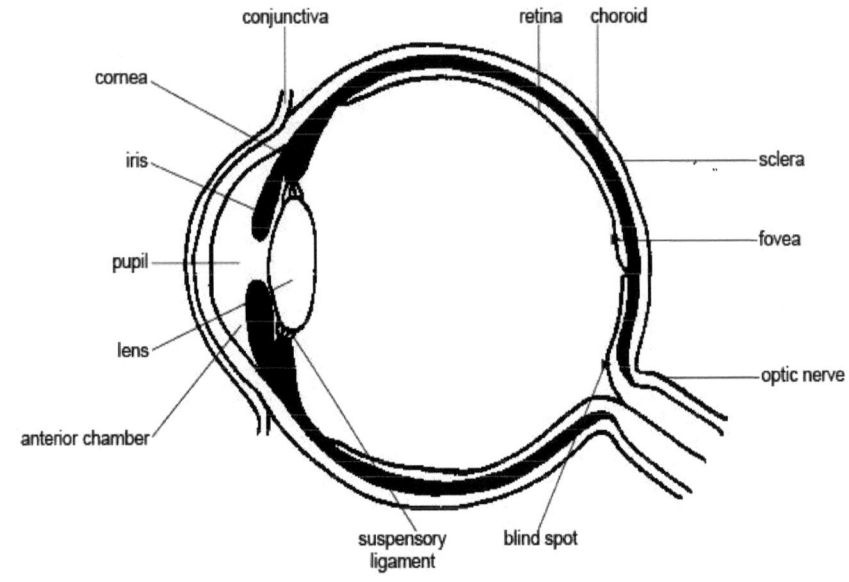

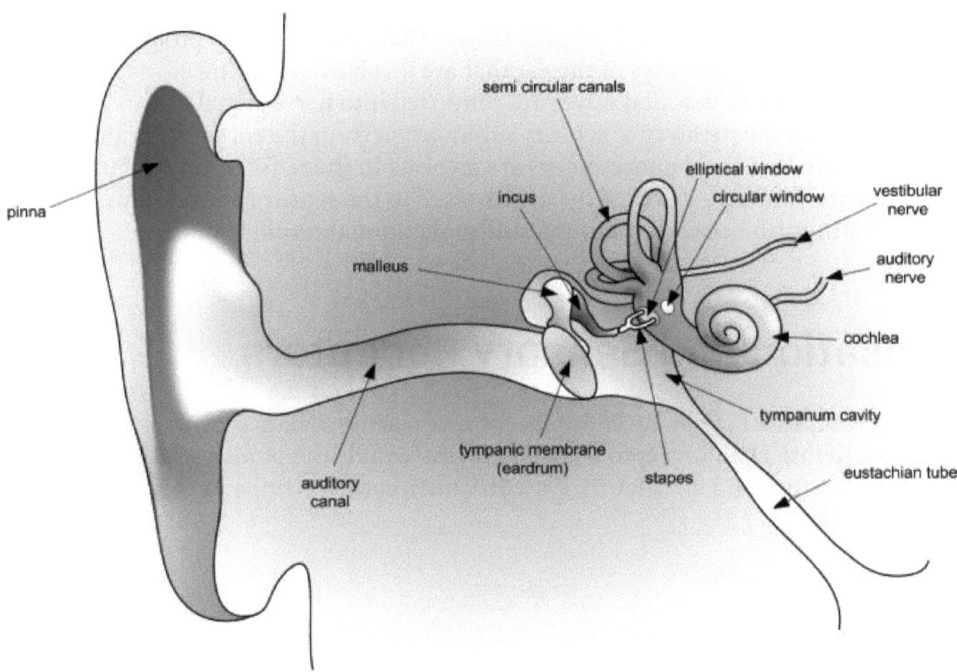

FIGURE 7.3 Human Eye and Ear

Figure 7.3 shows two of the special sensory organs—the eye and the ear.

Vision (Photoreceptors and Light)

The human eye can detect visible light. In the eye, rods and cones are two types of photoreceptor receptor cells found on the retina (at the back of the eye). The rods function in low light intensity whereas the cones (red, green, and blue) detect color, and function in higher light intensity. Light reaching a rod causes the breakdown of a chemical called rhodopsin (a compound formed from Vitamin A), which then causes an action potential that can be transmitted via the optic nerve to the occipital cortex of the brain.

Hearing

Hearing involves sound waves in the air being converted into vibrations, which ultimately cause movement of hair cells in the cochlea (inner ear). They are then converted into action potentials and sent through the auditory nerve to the auditory cortex in the temporal lobe of the brain.

Orientation and Gravity

Orientation and gravity are detected by the semicircular canals within the inner ear. Like hearing, hair cells within these canals provide the sense of equilibrium. Little "rocks in your head" here move in response to gravity, providing sensory information about gravity and acceleration to portions of the brain that deal with equilibrium and balance.

Key Points

Each of the various sensory receptors respond to a specific stimulus, important to maintaining homeostasis.

- There are chemoreceptors (which respond to chemical substances), pain receptors (called nociceptors—a chemoreceptor that responds to chemicals released by damaged tissues, and produces the sensation of pain), photoreceptors (which respond to light), mechanoreceptors (which respond to mechanical pressure), and thermoreceptors (which respond to temperature changes).
- This information leads to reflexes and actions that maintain homeostasis.
- Senses with receptors associated with skin, muscles, joints, and viscera ("insides") are called somatic senses. They send nerve impulses through the spinal cord and up to the cerebral cortex.
- Proprioceptors are mechanoreceptors that maintain muscle tone, equilibrium, and posture, allowing for balance and body position to be maintained (proprioception).
- The dermis of the skin contains cutaneous receptors that sense touch, pressure, pain, and temperature. Temperature receptors and pain receptors are free nerve endings.
- Many organs have pain receptors (nociceptors), sensitive to chemicals released by damaged tissues.

- Taste and smell are chemical senses that are sensitive to molecules in food and air (tastants and odorants). Other chemoreceptors, found in the carotid arteries and aorta, are sensitive to the pH of the blood.
- The taste buds contain taste cells (for sweet, sour, salty, bitter, and umami) that stimulate sensory nerve fibers.
- Almost all of what perceived as "taste" is actually due to smell. Olfactory cells in the roof of the nasal cavity detect odor molecules.
- Vision depends on the eyes and the brain.
- The eye has three layers: the outer layer, the sclera (white of the eye and cornea); the middle layer (choroid), which absorbs light rays; and the rods and cones, which are located in the retina at the back of the eye.
- The cornea and the lens are responsible for allowing light rays to focus on the retina.
- The pathway for vision goes to the occipital cortex after light has stimulated photoreceptors in the retina.
- The rods function in dim light and at night, while the cones function in bright light (color vision). Breakdown of rhodopsin (which uses a form of vitamin A) in rods stimulates the nerve signals.
- There are three types of cones (red, green, and blue).
- There are no rods or cones where the optic nerve exits the retina (blind spot).
- Light entering the eye through the pupil goes through the lens. By changing shape, the lens then sends the rays of light to the retina where rods and cones are stimulated to send a signal through the optic nerve to the occipital cortex for processing of the image.
- Nearsightedness and farsightedness are caused by the lens not being able to focus objects properly on the retina. Astigmatism is caused by an uneven cornea. Color blindness (X-linked) and a misshapen lens are other abnormalities of the eyes.
- The ear has two sensory functions: hearing and balance. The sensory receptors for both consist of hair cells with a type of cilia (mechanoreceptors).
- The outer ear is the pinna and auditory canal, which sends sound waves to the tympanic membrane.
- The middle ear begins with the tympanic membrane and contains three bones (malleus, incus, and stapes). The malleus is stimulated by sound waves hitting the tympanic membrane; the stapes then transmits those pressure signals to the oval window.
- The fluid-filled inner ear contains the cochlea (hearing), the vestibule (balance), and the semicircular canals (balance).
- Vibrations set up waves within the cochlea, which contains hair cells embedded in a membrane. When the stereocilia of the hair cells bend, nerve signals go through the cochlear nerve and are then carried to the auditory cortex of the brain, where they are interpreted as sound.
- The vestibular nerve and proprioceptors are necessary for maintaining equilibrium.
- Rotational equilibrium depends on the stimulation of hair cells embedded in semicircular canals, each of which responds to head rotation in a different direction (x-, y-, and z-axes).
- Gravitational (acceleration) equilibrium depends on stimulation of hair cells on a membrane with calcium carbonate granules (otoliths, or ear rocks) within the utricle and the saccule of the inner ear. When the head bends forward/back ("yes") or left/right ("no"), it displaces the otoliths, causing the stereocilia to move. The direction of movement causes nerve impulses to indicate the type of movement. Continuous stimulation of the stereocilia causes motion sickness.

Figure Credits

Fig. 7.3a: Copyright © by Sunshineconnelly (CC BY 3.0) at https://commons.wikimedia.org/wiki/File:Anatomy_and_physiology_of_animals_Structure_of_the_eye.jpg.
Fig. 7.3b: Source: https://commons.wikimedia.org/wiki/File:HumanEar.jpg.

Hormonally Speaking— The Endocrine System

Objectives

- **Identify** the organs and glands of the endocrine system.
- **Compare** the actions of peptide and steroid hormones.
- **Explain** the role of the hypothalamus in the endocrine system.
- **List** the hormones produced by the anterior and posterior pituitary glands and provide a function for each.
- **List** the hormones produced by the thyroid and parathyroid glands and provide a function for each.
- **Describe** the negative feedback mechanism involved in the maintenance of blood calcium homeostasis.
- **List** the hormones produced by the adrenal medulla and adrenal cortex and provide a function for each.
- **Explain** how the adrenal cortex is involved in the stress response.
- **List** the hormones produced by the pancreas and provide a function for each.
- **Describe** how the pancreatic hormones help maintain blood glucose homeostasis.
- **Distinguish** between type 1 and type 2 diabetes mellitus.
- **List** the hormones produced by the sex organs, thymus, and pineal gland.
- **List** hormones produced by glands and organs outside of the endocrine system.

The nervous system coordinates rapid and precise responses to stimuli using action potentials. The endocrine system maintains homeostasis and long-term control using chemical signals. The endocrine system works in parallel with the nervous system to maintain homeostasis.

Hormones

The endocrine system is a collection of glands that secrete chemical messages called hormones. These chemicals travel through the blood to a target organ. These targets have special receptors to accept the hormone and act on their message. Hormones may be grouped into different classes: steroids (lipids derived from cholesterol, like testosterone, the male sex hormone, and estrogens, the female sex hormone), peptides (short chains of amino acids produced by the pituitary, parathyroid, heart, stomach, liver, and kidneys), and the amines, a group of single amino acid hormones, derived from tyrosine, are produced by the thyroid gland and the adrenal medulla).

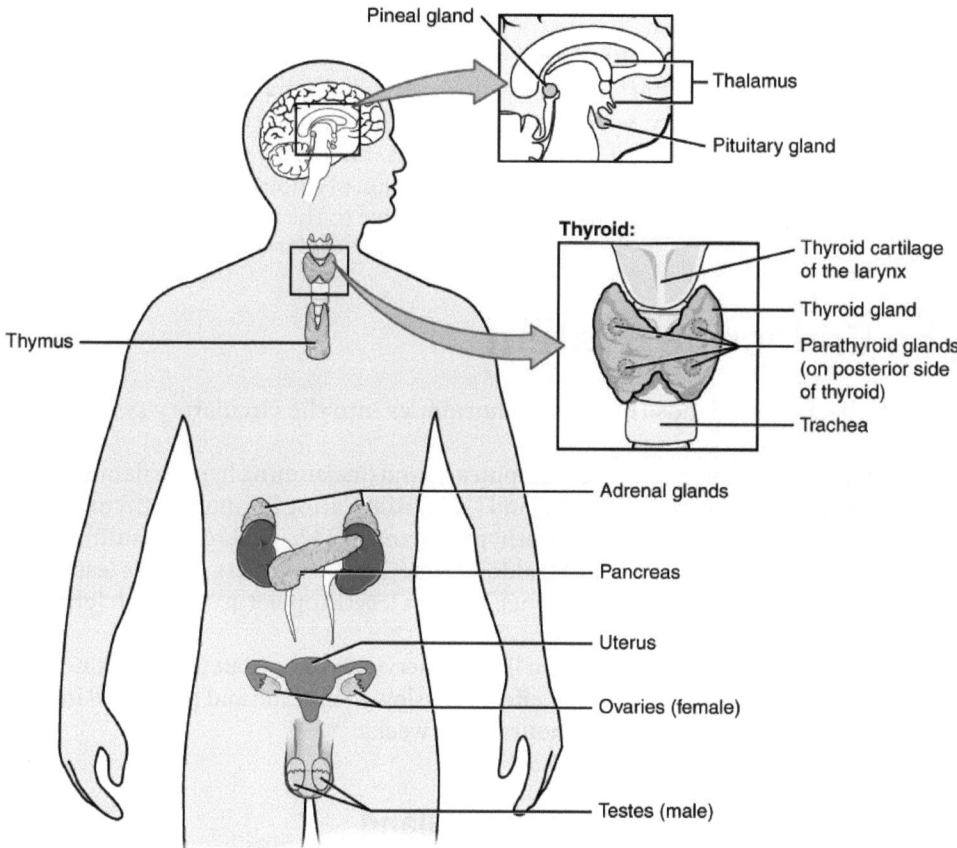

FIGURE 8.1. The Endocrine System

Figure 8.1 shows the various endocrine glands and their locations in the human body.

Endocrine Systems and Feedback Cycles

The endocrine system uses cycles and negative feedback to regulate physiological functions. Negative feedback regulates the secretion of almost every hormone (a major exception is oxytocin from the posterior pituitary).

Hormone Action and Potential Problems

The endocrine system releases hormones that activate actions in target cells. Receptors on the target cells bind specifically to one specific hormone. The binding hormone causes the cellular response to the hormone. Nonsteroidal hormones (water soluble, most of the peptides) do not enter the cell (because they are polar, hydrophilic molecules) but bind to membrane receptors, where they activate a second messenger system inside the cell. Second messengers activate other intracellular substances to produce the final cell response. The steroid hormones pass through the plasma membrane and form a hormone-receptor complex, which binds to DNA in the nucleus, activating specific genes to increase production of particular proteins. Endocrine-related problems may involve overproduction of a hormone, underproduction of a hormone, or nonfunctional receptors that cause target cells to become insensitive to the hormones.

The Endocrine System

- Collection of glands that secrete hormones into the circulatory system to be carried to a target organ.
- **Major endocrine glands include:** pineal gland (melatonin), hypothalamus (_RH, releasing hormones), pituitary gland (_SH, stimulating hormones), thyroid gland (thyroxine—T_4, calcitonin), parathyroid gland (PTH), pancreas (insulin, glucagon), adrenal glands (cortisol, aldosterone, sex hormones), ovaries (estrogen, progesterone), testes (testosterone), kidneys (erythropoietin), stomach (ghrelin), and adipose tissue (leptin).
- It's an information signal system like the nervous system, but unlike the nervous system, the endocrine system's effects are slow to initiate, and prolonged in their response, lasting from a few hours up to weeks.

The Hypothalamus and Pituitary Gland

The pituitary gland is found in a small bony recess in the skull, where it communicates with the hypothalamus, which controls release of the pituitary hormones. Seven different hypothalamic hormones can be released to cause the pituitary to release eight of its different hormones. The pituitary gland has two lobes: the anterior and posterior lobes.

Growth hormone (GH) is a peptide hormone from the anterior pituitary and is essential for growth. GH-releasing hormone, from the hypothalamus, stimulates release of GH. GH-inhibiting hormone, also from the hypothalamus, suppresses the release of GH. Cells under the influence of GH increase in size (hypertrophy) and number (hyperplasia). GH also causes an increase in bone length and density. During adolescence, sex hormones cause replacement of cartilage by bone, halting further bone growth even though GH is still present. Too little GH can cause dwarfism, and too much GH can cause gigantism. Increased GH in adults causes acromegaly.

Hypothalamic receptors monitor blood levels of hormones, such as those from the thyroid gland (T_3 and T_4). Low blood levels of either of these cause release of thyrotropin-releasing hormone (TRH) from the hypothalamus. TRH then causes the release of thyroid-stimulating hormone (TSH) from the anterior pituitary. TSH travels to the thyroid, where it promotes production of thyroid hormones (T_3 and T_4). They will then affect metabolic rates and temperatures in the body.

Gonadotropins (follicle-stimulating hormone, FSH, and luteinizing hormone, LH) and prolactin are also secreted by the anterior pituitary in response to the release of gonadotropin-releasing hormone (GnRH) from the hypothalamus. These gonadotropins affect the gonads by stimulating gamete formation (ova and sperm) and production of sex hormones (estrogens and testosterone). Prolactin is secreted near the end of pregnancy and prepares the breasts for milk production (lactation).

The posterior pituitary stores and releases hormones into the blood. Antidiuretic hormone (ADH, vasopressin) and oxytocin are both produced in the hypothalamus and then transported via axons to the posterior pituitary, where they are stored until needed. ADH controls water balance through the kidneys and also affects blood pressure. Oxytocin is a small peptide hormone that stimulates uterine contractions during childbirth, and milk ejection during nursing.

The Other Endocrine Organs

The Thyroid Gland

The thyroid gland is found in the neck. Thyroid-stimulating hormone (TSH) from the anterior pituitary causes production and release of the thyroid hormones T_4 and T_3. Almost all body cells have receptors for, and are targets of, the thyroid hormones, which increase the overall metabolic rate, and regulate growth and development. Calcitonin, which helps to regulate calcium homeostasis by decreasing Ca^{++} when its level increases, is also produced by cells of the thyroid.

The Parathyroid Glands

These four little glands are located on the back of the thyroid gland and increase calcium (Ca^{++})—opposing the action of calcitonin from the thyroid.

The Pancreas

The pancreas contains exocrine cells that secrete digestive enzymes into the small intestine, and clusters of endocrine cells (the pancreatic islets) that produce the hormones insulin and glucagon to help regulate blood glucose homeostasis. After a meal, blood glucose levels rise, and insulin is released. This causes cells to take up glucose from the blood through the glucose transporters produced in response to insulin. The liver and skeletal muscle cells can store glucose as glycogen, or use it directly to generate energy (ATP). As glucose levels in the blood fall, insulin production is shut off. Glucagon causes the breakdown of glycogen into glucose in the liver to maintain glucose levels within a homeostatic range. Glucagon release stops when blood glucose levels again are in the normal range.

Diabetes mellitus results from inadequate levels of insulin, or insulin insensitivity (cells don't respond to insulin even if it is present in the blood). Type I diabetes is due to lack of insulin secretion. Type II diabetes usually develops when there is a loss of response to insulin by the cells. Diabetes causes multiple "-opathies" in the eyes (retinopathy), circulatory system (vasculopathy), nervous system (neuropathy), and kidneys (nephropathy).

The Adrenal Glands

Each kidney has an adrenal gland sitting above it. The adrenal gland has an inner medulla and an outer cortex. The adrenal medulla synthesizes amine hormones (epinephrine, norepinephrine), and the cortex secretes steroid hormones (glucocorticoids: cortisol that raises glucose levels and suppresses the immune response as well as inhibiting the inflammatory response; mineralocorticoids: aldosterone that maintains electrolyte balance; and a small amount of the sex hormones).

Key Points

Endocrine glands produce hormones. They are involved in regulating other organs in the quest to maintain homeostasis throughout the body by working closely with the nervous system. The endocrine system provides a slow prolonged response, whereas the nervous system responds quickly to stimuli.

- Endocrine glands secrete their hormones directly into the blood.
- Only cells with the right receptors (target cells) can respond to a specific hormone.
- Peptide hormones are peptides, proteins, glycoproteins, and modified amino acids.
- Steroid hormones are all derived from cholesterol.
- Peptide hormones bind to a hormone receptor located on the cell's membrane, leading to activation of a second messenger enzyme cascade.
- Steroid hormones enter the nucleus to activate DNA, stimulating protein synthesis.
- Thyroid hormones act like steroid hormones, but they are not steroids.
- The hypothalamus controls secretions of the pituitary gland.
- The pituitary gland has an anterior and a posterior component.
- The posterior pituitary secretes antidiuretic hormone (ADH, vasopressin) and oxytocin, both of which are produced in the hypothalamus but are stored in the posterior pituitary.
- Antidiuretic hormone (ADH) helps to regulate the water–salt balance of the blood.
- Lack of ADH causes diabetes insipidus.
- Oxytocin stimulates uterine contractions during labor and childbirth. After that, it stimulates milk letdown during nursing.
- Oxytocin is controlled by positive feedback.
- The anterior pituitary produces several hormones.
- Thyroid-stimulating hormone (TSH) stimulates the thyroid.
- Adrenocorticotrophic hormone (ACTH) stimulates the adrenal cortex.
- FSH and LH (gonadotropic hormones) stimulate the gonads.
- Prolactin causes mammary glands to produce milk.
- Growth hormone promotes bone growth.
- The hypothalamus regulates release of anterior pituitary hormones through secretion of releasing or inhibiting hormones (TRH, CRH, GnRH).
- Pituitary dwarfism is caused by too little growth hormone in childhood, and gigantism is caused by too much growth hormone during childhood. Too much growth hormone in adulthood causes acromegaly.

- The thyroid gland is a gland located in the neck with the parathyroid glands "hidden" in the back of the thyroid gland.
- The thyroid gland produces the hormones thyroxine (T_4) and triiodothyronine (T_3).
- Thyroid hormones generally increase metabolism.
- The thyroid gland also produces calcitonin to lower blood calcium levels.
- The parathyroid glands produce parathyroid hormone (PTH), which causes a rise in blood calcium by causing osteoclasts to breakdown bone, causes the kidneys to reabsorb more calcium, and also activates the conversion of vitamin D_2 to D_3, which helps the body absorb more calcium from ingested foodstuff.
- The pancreatic islet cells produce and secrete insulin, which lowers blood glucose. The islet cells also produce and secrete glucagon, which raises blood glucose. These two hormones work together maintain glucose homeostasis.
- Diabetes mellitus (DM) is a disease in which body cells can't take up glucose from the blood.
- Type 1 DM (insulin-dependent diabetes) is due to lack of pancreatic insulin. Patients have to inject insulin daily.
- Type 2 DM (noninsulin-dependent diabetes) is more common than type 1. It is caused by insulin insensitivity (cells don't recognize or respond to the high levels of circulating insulin).
- The adrenal glands are located on top of each kidney. They have an inner medulla and an outer cortex. The adrenal medulla is basically part of the autonomic nervous system and produces epinephrine (adrenaline) to respond to acute stress (fight/flight/fright). The adrenal cortex is regulated by hypothalamic corticotropic-releasing hormone (CRH) and pituitary adrenocorticotrophic hormone (ACTH). These allow the adrenal cortex to respond to chronic stress, hypoglycemia (cortisol stimulates gluconeogenesis to produce more glucose from proteins), immunologic issues (immunosuppression), and sodium balance, via aldosterone.
- Aldosterone causes the kidneys to reabsorb sodium ions (Na^+) and excrete potassium ions (K^+).
- The renin-angiotensin-aldosterone system causes the adrenal cortex to release aldosterone and produces vasoconstriction to raise blood pressure.
- The adrenal cortex also secretes small amounts of the sex hormones, androgens and estrogens.
- Addison's disease develops from low blood levels of glucocorticoids.
- Cushing's syndrome develops from excessive levels of glucocorticoids.
- The gonads are the testes in males and ovaries in females. They produce sex hormones (testosterone in males and estrogen and progesterone in females).
- Miscellaneous hormones:

  - Thymosin is produced by the thymus to stimulate T lymphocyte production and maturation.
  - Melatonin (for circadian rhythms) is produced by the pineal gland.
  - Erythropoietin from the kidneys stimulates red blood cell production in the red bone marrow.
  - Leptin is produced by adipose tissue, signaling to the hypothalamus the sensation of satiety.

> Gluconeogenesis: new glucose formation, often from proteins or other carbon nutrients (NOT from fat)

Figure Credit

Fig. 8.1: Copyright © by OpenStax (CC BY 3.0) at https://commons.wikimedia.org/wiki/File:1801_The_Endocrine_System.jpg.

Heart-y Functions and Blood

LESSON 9.1

Objectives

- **Identify** the two components of the cardiovascular system and their functions.
- **Identify** the structures and chambers of the human heart.
- **Describe** the flow of blood through the human heart.
- **Explain** the internal and external controls of the heartbeat.
- **Distinguish** between systolic and diastolic pressure.
- **Compare** blood flow and oxygenation in the pulmonary and systemic circuits.
- **Explain** the underlying causes of cardiovascular disease in humans.

By looking at the functions and interrelationships between the heart, blood vessels, and lungs, you will gain an appreciation of this amazing interrelated system. You will be introduced to the pumping mechanisms of the heart and vessels, as well as the composition and functions of the blood. You will see the path of a drop of blood through the heart, to the lungs, and then through the rest of the body (from the aorta out to the body, or through the pulmonary arteries to the lungs and back to the heart). Finally, you will see what is called the cardiac cycle.

The Heart

The heart is a muscular structure, composed of specialized cardiac muscle cells that allow for rhythmic contractions to pump blood through its four chambers, to the lungs (pulmonary circuit), or to the rest of the body (systemic circuit). An atrioventricular (AV) valve separates the atria (two receiving chambers) from the ventricles (two discharging chambers). The mitral valve is the name given to the bicuspid valve between the left atrium and ventricle. A semilunar valve separates each ventricle from its connecting artery (either the pulmonary artery on the right side, or the aorta on the left side).

A typical heart beat is approximately 70 times per minute. The human heart goes through a specific cycle of events, known as the cardiac cycle. This cardiac cycle involves cycling between systole (contraction) and diastole (relaxation). The atria contract while the ventricles relax, and the atria relax when the ventricles contract. Valves in the heart open and close during this cycle. This contraction cycle is controlled by special electrical "pacemakers" within the heart itself: the SA node (sinoatrial node) initiates heartbeat at about 60 to 80 beats per minute (BPM). The AV node (atrioventricular node) receives a charge from the SA node, and then causes the ventricles to contract. The SA node is typically the pacemaker. Heartbeat is also influenced by the autonomic nervous system (increased rate of contractions by the sympathetic, and decreased rate of contractions by the parasympathetic nervous system).

Figure 9.1 shows the human cardiovascular system—a system of the heart, involved in pumping, and a network of vessels, responsible for the flow of blood, nutrients, oxygen

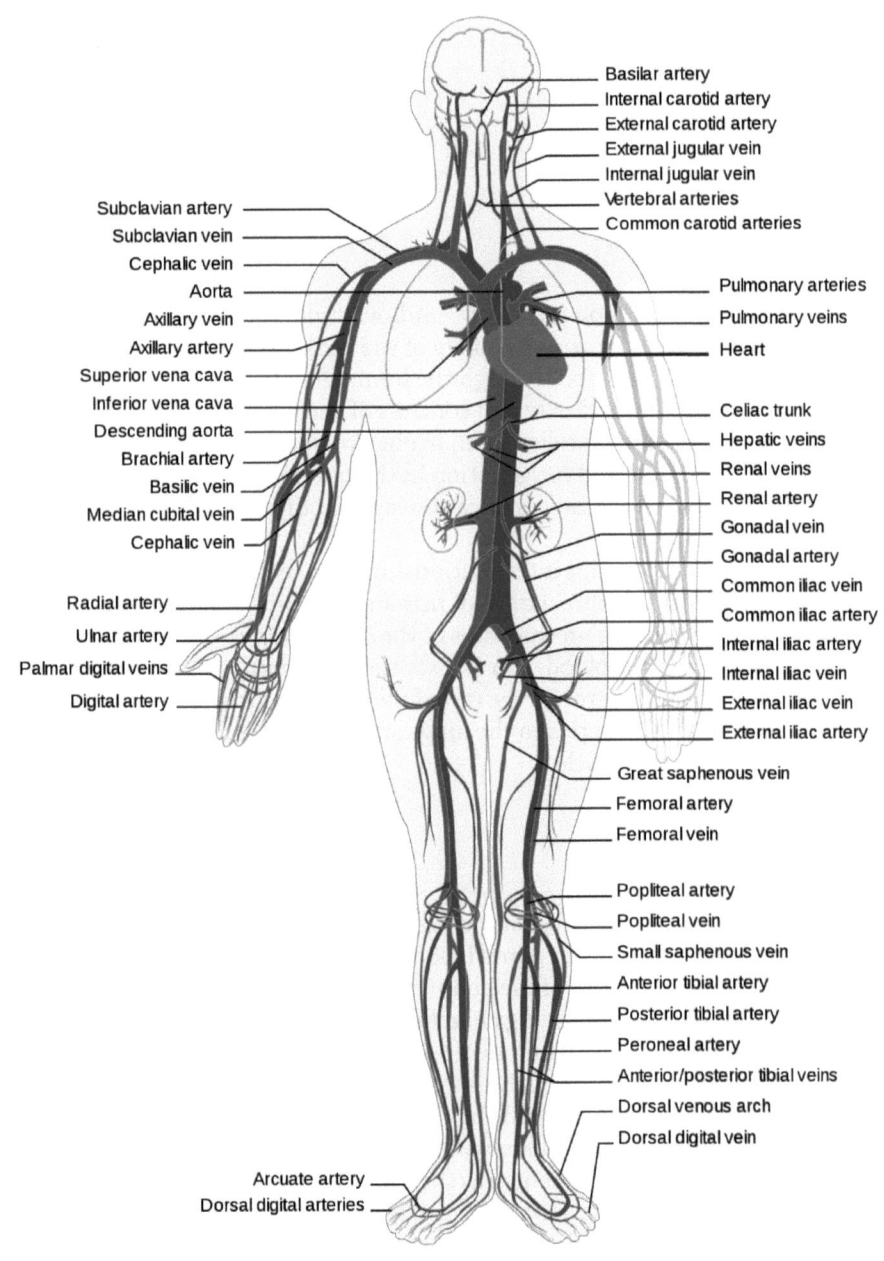

FIGURE 9.1. Cardiovascular System

and other gases, waste products, and hormones to and from cells. Without the circulatory system, the body would not be able to fight disease or maintain a stable internal environment (such as proper temperature and pH; i.e., homeostasis).

Blood flows through the heart: blood enters the heart through the venae cavae (superior and inferior) into the right atrium (simultaneously, oxygenated blood from the lungs flows from the pulmonary vein into the left atrium); then the muscles of both atria contract, forcing blood down through each AV valve into the ventricles; this is followed by ventricular contraction which forces blood through the semilunar valves out the pulmonary artery (right side) or aorta (left side). The sound of the heart valves opening and closing produces the classically described "lub-dub" sound.

An electrocardiogram (ECG, EKG) measures changes in electrical activity through the heart. The graphed changes in electrical activity produce the typical P-Q-R-S-T waves. The P wave represents atrial depolarization; the QRS complex represents ventricular depolarization; and the T wave represents ventricular repolarization. Where does atria repolarization occur? (It is hidden behind the QRS complex!)

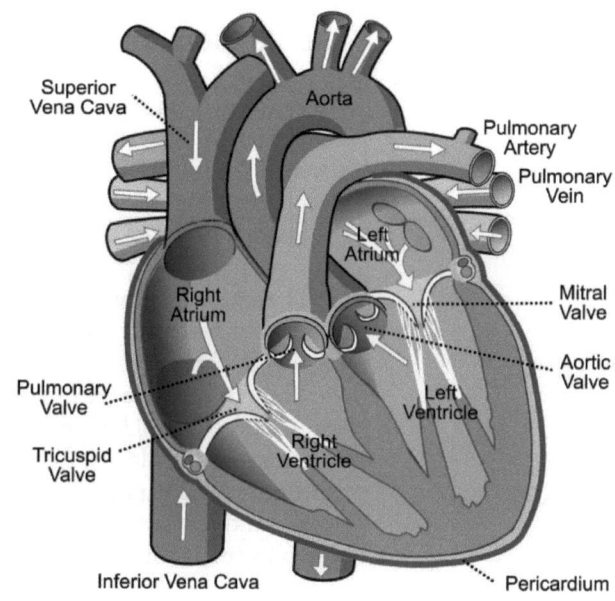

FIGURE 9.2. The Human Heart, Its Chambers, and the Vessels That Enter and Leave. Arrows Indicate the Flow of Blood through the Heart.

The Circulatory System

Because many of your cells are not in contact with a source of oxygen and nutrients, you have a circulatory system to transport nutrients and oxygen to the cells, and transport carbon dioxide and metabolic waste products away from the cells. The circulatory system includes blood (connective tissue of plasma and cells, especially red blood cells, which contain hemoglobin, a molecule that transports oxygen), the heart (a muscular pump which moves the blood), and blood vessels (arteries, arterioles, capillaries, venules, and veins that transport the blood to the cells).

Your cardiovascular system includes the heart and blood vessels. The upper chambers of the heart, the atria, are the entry point for the blood to enter the heart. Passing through a different valve on each side, blood flows to the lower ventricles. Contraction of the ventricles forces the blood from the heart through either the pulmonary artery or the aorta. The heart muscle is composed of specialized electrical cardiac muscle cells.

The Vessels of the Vascular System

There are two main routes for circulation: pulmonary (to and from the lungs) and systemic (to and from the body). Pulmonary arteries carry deoxygenated blood from the heart to the lungs. In the lungs, gas exchange occurs between the capillaries and the alveoli (little air sacs). Pulmonary veins carry oxygenated blood back from the lungs to the heart. The aorta is the main artery of the systemic circuit. The venae cavae are the main return veins from body to the heart in the systemic circuit.

The arteries carry blood away from the heart, assisted by their smooth muscle fibers that can either contract or relax to respond to different physiological conditions. The aorta is the main artery leaving the heart. The pulmonary artery is the only artery in the body that carries deoxygenated blood. It carries this deoxygenated blood to the lungs, where gas exchange occurs (carbon dioxide diffuses out of the blood, and oxygen diffuses into the blood). Arterioles, small, muscular vessels, connect the arteries with capillaries. The capillaries are very thin-walled vessels that allow for the exchange of oxygen and nutrients with the cells in exchange for carbon dioxide and waste products, which in turn will be removed from the body. Interestingly, blushing is one manifestation of blood flow into capillaries. Blood then leaves the capillaries and flows into progressively larger vessels called venules that lead the blood into the veins. The veins carry blood to the heart. Other than the pulmonary veins, which carry oxygenated blood, the veins all carry deoxygenated blood. The veins, which have valves, use skeletal muscle contractions to move blood back towards the heart. Blood pressure, measured in mm of mercury, is considered to be normal when systolic pressure (ventricular contraction) is approximately 120 mm, whereas diastolic pressure (ventricular relaxation) is approximately 80 mm of mercury (Hg). Blood pressure is sensed by special receptors in the certain arteries. Transmitted neural signals from these receptors to the medulla help the medulla regulate blood pressure.

Diseases of the Heart and Cardiovascular System (Mostly Due to Hypertension or Atherosclerosis)

The coronary arteries are the first branch off the aorta and "feed" the heart itself. Blocked coronary artery blood flow can cause death of cardiac muscle, leading to a myocardial infarct (heart attack). This blockage of the coronary arteries is usually the result of gradual buildup of plaque within the inner wall of the coronary artery. Occasional chest pain is called angina pectoralis. Angina indicates lack of oxygen delivery when the demands are greater than the ability to deliver adequate oxygen to the heart, and indicates that a heart attack may occur in the future. Unfortunately, cardiac muscle cells that die cannot normally be replaced.

Hypertension, or high blood pressure (a silent killer—no symptoms), occurs when blood pressure is consistently above 130/90. Causes may include a genetic predisposition made worse by stress, obesity, high salt intake, and smoking. Atherosclerosis is a hardening of the arteries/arterioles due to the buildup of fatty plaque within the vessel wall. This is usually the result of an initial small nick in the vessel wall. During repair of the damaged wall, certain substances floating by may get trapped in the wall and ultimately decrease the vessel's diameter. This severely restricts the flow of blood to the tissues that need oxygen and nutrients. Some of the trapped substances include platelets, calcium, and fat/cholesterol.

Key Points

The cardiovascular system consists of the heart and the blood vessels through which blood flows.

- The five types of blood vessels are arteries, arterioles, capillaries, venules, and veins.
- Arteries and arterioles transport blood away from the heart. They both have muscular elastic tissue that allows them to help regulate blood pressure via constriction or dilation of the arterioles.
- Capillaries have walls that are one cell thick to allow for easy exchange of substances between cells and blood.
- Venules drain into veins to return blood to the heart. Veins have much less smooth muscle than arteries and arterioles, but often have valves to prevent backflow of blood.
- The right and left atria empty into the right and left ventricles, respectively. Atrioventricular (AV) valves exist between atria and ventricles, and semilunar valves exist between the ventricles and the output arteries (pulmonary artery going to the lungs, aorta going to the rest of the body).
- The heart receives oxygen and nutrients from the coronary arteries, the first branch off the aorta.
- The path of blood from the body through the heart starts with return of blood from the body through superior and inferior venae cavae and then through the right atrium, right ventricle, pulmonary arteries, and lungs before blood returns to the heart via pulmonary veins, then through the left atrium, left ventricle, and then out to the body through the aorta.
- Each heartbeat is accomplished via what is known as the cardiac cycle.
- Systole is the contraction of atria and ventricles.
- Diastole refers to relaxation of atria and ventricles.
- The heart sounds, "lub-dub," are from closing of the atrioventricular valves (lub), followed by the closing of the semilunar valves (dub).
- The sinoatrial (SA) node (cardiac pacemaker) beats automatically and causes atrial contraction before the signal travels down towards the ventricles.
- The atrioventricular (AV) node next sends the signal through the ventricular conduction system (from the AV node to the atrioventricular bundle [of His], then bundle branches, then Purkinje fibers) and causes ventricular contraction.
- A cardiovascular center in the medulla oblongata can also affect the heartbeat via the parasympathetic nervous system (slows heart rate) and sympathetic nervous system (increases heart rate).
- An electrocardiogram (ECG/EKG) records electrical changes within the myocardium during a cardiac cycle; P wave = atrial depolarization; QRS complex = ventricular depolarization; T wave = ventricular repolarization. So, where is atrial repolarization?
- Pumping of the heart allows for homeostasis by creating sufficient pressure to send blood throughout the body (especially to the brain).

- Systolic and diastolic pressure can be measured with a sphygmomanometer (blood pressure cuff). Normal blood pressure is about 120/80 (systolic/diastolic). High blood pressure is referred to as **hypertension**, and low blood pressure is referred to as **hypotension**.
- Blood flows in two circuits—pulmonary and systemic.
- The pulmonary circuit has arteries that send blood from the right ventricle to the lungs, allowing for the exchange of carbon dioxide and oxygen. Pulmonary veins return blood to the left atrium. Pulmonary arteries carry oxygen-poor blood, while pulmonary veins carry oxygen-rich blood (the only place in the body where that occurs).
- The aorta sends blood out from the heart to the rest of the body.
- The largest veins of the body (superior and inferior venae cavae) return blood to the right heart.
- The hepatic portal system takes blood directly from the digestive tract to the liver.
- After blood has been filtered through the liver, it is returned to the heart via the inferior vena cava.
- Hypertension ("the silent killer") can lead to heart disease, such as atherosclerosis (accumulation of plaque in the linings of arteries). If any of these plaques break off and get stuck in a small vessel (thromboembolism), it can be deadly (deep vein thrombosis [DVT] or pulmonary embolism [PE]).
- If a blood clot gets stuck in a cerebral vessel (or a cerebral vessel bursts), it is known as a stroke and results in part of the brain dying.
- If a blood clot clogs up a coronary vessel, a heart attack (myocardial infarct [MI]) occurs. An aneurysm is ballooning of a vessel, and if a major vessel bursts, death may occur.
- Heart failure occurs when the heart is no longer capable of pumping enough blood to meet the needs of the body.
- Blood allows for the transport of gases (O_2 and CO_2) and nutrients to cells as well as removal of waste products.
- Blood also provides defense for the body against pathogens and contains substances to prevent blood loss.
- Blood contains cells (red and white), cell fragments (platelets), and plasma.
- Plasma is liquid (mostly water), but it also contains salts, small molecules, hormones, and plasma proteins.

Figure Credits

Fig. 9.1: Source: https://commons.wikimedia.org/wiki/File:Circulatory_System_en.svg.
Fig. 9.2: Copyright © by Wapcaplet (CC BY-SA 3.0) at https://commons.wikimedia.org/wiki/File:Diagram_of_the_human_heart_(cropped).svg.

Blood

LESSON 9.2

Objectives

- **List** the functions of blood in the human body.
- **Compare** the composition of formed elements and plasma in the blood.
- **Explain** the role of hemoglobin in gas transport.
- **Compare** the transport of oxygen and carbon dioxide by red blood cells.
- **Summarize** the role of erythropoietin in red blood cell production.
- **Explain** the function of white blood cells in the body.
- **Explain** how blood clotting relates to homeostasis.
- **Explain** what determines blood types in humans.
- **Summarize** the role of Rh factor in hemolytic disease of the newborn.

The Blood

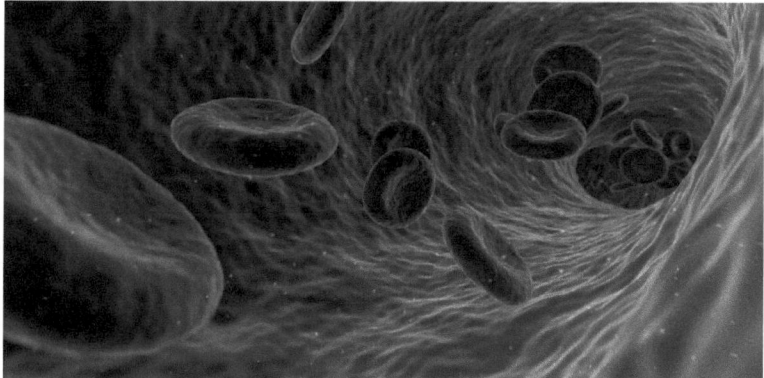

FIGURE 9.3. An Electron Micrograph of Blood. Note the Unique Biconcave Disc Shape of the Red Blood Cells.

Three Types of Formed Elements

- **Erythrocytes**, or **red blood cells**, carry oxygen on the protein hemoglobin.
- **Leukocytes**, or **white blood cells**, are involved in defending the body against foreign invaders.
- **Platelets** are cell fragments involved in blood clotting (also called thrombocytes).

White blood cells are characterized by presence or absence of **granules** in their cytoplasm.

Three Types of Granulocytes

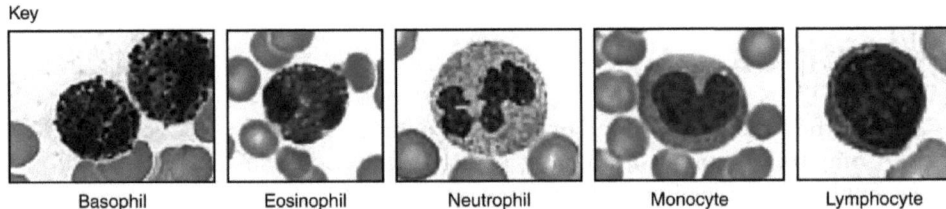

FIGURE 9.4. The Five Different Types of White Blood Cells

- **Neutrophils:** Most abundant white blood cell. Predominant cells in pus, accounts for its whitish appearance. Respond quickly following tissue injury. Hallmark of acute inflammation.
- **Basophils:** Least common granulocyte. When activated, release histamine and other inflammatory chemicals.
- **Eosinophils:** Main effector cells in allergic responses and asthma. Also fight helminth (worm) infections.

Neutrophils and eosinophils can **phagocytize** pathogens.

> Phagocytize: process of ingestion whereby the cell membrane reaches out to engulf particles for processing

Two Types of Agranulocytes

- **Lymphocytes:** intimately involved in specific immunity (third line of immune defense)
- **Monocytes:** leave the blood and mature into **macrophages** (phagocytic cells involved in the second line of immune defense)

Plasma is the liquid portion of the blood. Blood is made up of liquid (plasma) and cellular or cell fragment components. Plasma makes up about 55 to 60 percent of the blood; cells and fragments make up the other 40 to 45 percent. Plasma is 90 percent water and 10 percent dissolved materials (proteins, glucose, ions, hormones, and gases). It assists in buffering pH (resists a change in pH), maintaining pH near 7.4. Plasma proteins help to transport large molecules that can't be transported without assistance and prevent edema by helping to keep water within the vessels.

Red blood cells (erythrocytes) are biconcave disks lacking a nucleus that are essentially bags of hemoglobin, whose function is to carry oxygen to cells. Red blood cells are continuously produced and released in the red bone marrow of long bones, ribs, skull, and vertebrae (in response to the kidney hormone erythropoietin). Erythrocytes live only about 120 days. They are then destroyed in the liver and spleen. Iron from the hemoglobin molecule is retained and reused by the red marrow. The liver converts the heme molecules to a water-soluble bilirubin, which is secreted in the bile to be excreted in the feces, or excreted in urine, and is responsible for the color of each.

White blood cells (leukocytes) are larger than erythrocytes and have a nucleus. They function in the immune system to protect the body from invading organisms or particles. The white blood cells make up less than 1 percent of the blood's volume. Like red blood cells, they come from stem cells in the bone marrow. There are five types: neutrophils, which phagocytose foreign substances; monocytes/macrophages, which are "big eaters" and help to process foreign particles to be dealt with by the immune system; lymphocytes, which help fight infection (T lymphocytes attack cells containing viruses or cancer, B lymphocytes produce antibodies); eosinophils and basophils are minor white blood cell components, and will be discussed with the immune system. White blood cells can squeeze through pores in the capillaries (diapedesis) and fight diseases.

Platelets are cell fragments from bone marrow cells called megakaryocytes and are involved with clotting. Platelets survive for 10 days before being removed by the liver and spleen. Platelets stick and adhere to cuts in blood vessels. They also activate clotting factors (special plasma proteins). A hemophiliac's blood cannot clot due to lack of one of these proteins (factor VIII, or factor IX). Supplying the correct proteins (clotting factors) has been a common method of treating hemophiliacs. Unfortunately, prior to current-day testing, it also led to accidental HIV transmission due to the use of transfusions and contaminated blood products. This even led to AIDS in some hemophiliacs.

Together, the blood vessels, platelets, and clotting factors maintain hemostasis (the homeostasis of preventing blood loss). When a vessel is injured, the initial reaction is vasoconstriction, as a smaller diameter vessel loses less blood. This, though, only lasts a short time. Tissue components from the damaged vessel activate platelets to clump and plug up the hole. Again, this only lasts a short time. Fortunately, the platelets can release a substance that starts a cascade of events by the clotting factors. Eventually a stable fibrin clot is formed. This will remain until the tissue has had a chance to repair itself, at which time the clot will be dissolved. Homeostasis!

Key Points

- Blood brings cells, oxygen, and nutrients, and removes carbon dioxide and waste products.

Red blood cells (erythrocytes) are small, biconcave disks that are basically a bag of hemoglobin (the oxygen-binding molecule).

- Carbon dioxide is mainly transported in the plasma as bicarbonate ions. These bicarbonate ions reach the lungs and are converted to carbon dioxide, which then diffuses out of the blood into the lungs to be exhaled.

- Red blood cells live only about 120 days, then are destroyed in the liver and the spleen.
- Heme iron can be saved and sent to the bone marrow. Breakdown of hemoglobin produces bilirubin (fat soluble until the liver makes it water soluble).
- Anemia is caused by having too few red blood cells and/or too little hemoglobin.

White blood cells (leukocytes) are large and nucleated (unlike erythrocytes). They function in immunity.

- Some white blood cells are phagocytes.
- Others white blood cells make antibodies.
- White blood cells are classified as granular or agranular.
- Neutrophils (or polymorphonuclear leukocytes) are granulocytes that engulf bacteria and debris through phagocytosis.
- Eosinophils are found during allergic reactions or parasitic infections.
- Basophils (which in tissues are called mast cells) release histamine (allergies).
- Agranular leukocytes include monocytes (which become macrophages in tissues), which phagocytose pathogens and cellular debris; and lymphocytes, the major agranulocyte, which may become either B cells or T cells, and are involved in specific immunity.

Platelets (thrombocytes) are fragments of megakaryocytes from the bone marrow.

- Platelets are involved in blood clotting.
- When tissues are damaged, the first response to prevent blood loss is vasoconstriction, but that doesn't last very long. Platelets are activated and plug up the damaged area.
- Finally, the clotting system is activated and after prothrombin is converted to thrombin, the thrombin converts fibrinogen to fibrin, forming a strong, stable clot. Now the damaged vessel can repair itself.
- Hemophilia is a clotting disorder resulting from deficiency of a clotting factor.
- A clot in a vessel is called a **thrombus**, but if it breaks free and travels, it is called an **embolus**.

Blood typing determines the ABO and Rh blood groups.

- The blood types A, B, AB, and O are determined by finding specific antigens (chemical structures) on the surface of red blood cells (RBCs).
- You may have antibodies in your plasma for the A and/or B antigen if that antigen is not found on your own RBCs.
- If these antibodies in your plasma interact with antigens on the surface of red blood cells, agglutination (clumping) will occur.
- The Rh (Rhesus) system involves a different antigen on the RBCs.
- Rh-positive means presence of the antigen, whereas Rh-negative means absence of the antigen.
- Rh factor is especially important in Rh-negative mothers who have Rh-positive babies because of the possibility of hemolytic disease of the newborn (HDNB).

Figure Credits

Fig. 9.3: Source: https://pixabay.com/illustrations/blood-cells-red-medical-medicine-1813410/.

Fig. 9.4: Copyright © by OpenStax (CC BY 3.0) at https://commons.wikimedia.org/wiki/File:1916_Leukocyte_Key.jpg.

Inspirational Breathing Functions

LESSON 10

Objectives

- **Summarize** the role of the respiratory system in the body.
- **Identify** the structures of the human respiratory system.
- **Explain** how the alveoli increase the efficiency of the respiratory system.
- **Distinguish** between inspiration and expiration.
- **Contrast** the processes of inspiration and expiration during ventilation.
- **Define** tidal volume, vital capacity, and residual volume in relation to ventilation.
- **Explain** the role of chemoreceptors and pH levels in regulating breathing rate.

The respiratory system includes the lungs, pathways connecting them to the outside, and structures in the chest involved with moving air in and out of the lungs. The lungs are the pair of organs of the respiratory system found within the thoracic cavity. The lungs are surrounded by two layers of membrane called pleura (parietal and visceral). Between each layer is a lubricating fluid that allows the lungs to expand and contract during respiration. Loss of this fluid, which usually allows for bonding of the two membranes, can result in a collapsed lung (pneumothorax). Air is inhaled into the lungs through the trachea and bronchi into sacs called alveoli. From the alveoli, oxygen diffuses across the capillary membrane into the blood. Carbon dioxide is transferred from the capillaries to these alveoli, and then exhaled. Let's look at the structure and function of the respiratory system.

The Respiratory System and Gas Exchange

Cellular metabolism is the process by which cells produce ATP. Sufficient oxygen is required for the aerobic metabolism by the Krebs cycle and electron transport chain to efficiently produce a large amount of ATP. Carbon dioxide is generated during cellular metabolism and must be removed. Humans have developed a system of respiratory structures that increase the surface area to aid in the exchange of gases between the body and the environment. It is the respiratory system that facilitates this exchange.

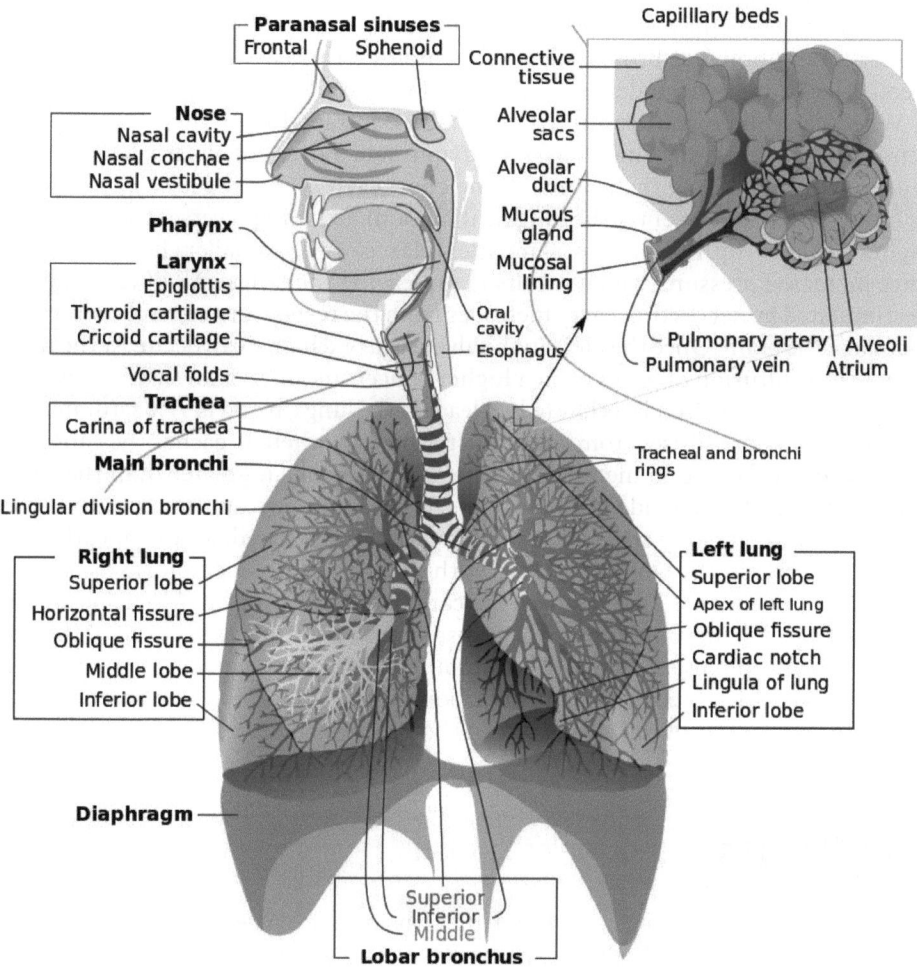

FIGURE 10.1. The Respiratory System

Figure 10.1, the respiratory system, has as its main function supplying the blood with oxygen (O_2) so that the blood can deliver it to all parts of the body. It also allows for the removal of carbon dioxide (CO_2) from the tissues.

Your respiratory structures are inside the body and are connected to the outside environment by a series of tubes. The trachea is the tube that carries air through bronchi and bronchioles to the lungs for gas exchange. This ultimately involves diffusion of the oxygen from the air into the blood and transportation of this oxygen to the cells of the body, followed by diffusion of the oxygen from the blood into cells. This process is carried out in reverse for carbon dioxide exchange.

Air enters the body through the nose, where it is warmed, filtered, and passed through to the pharynx (where the epiglottis prevents food from entering the respiratory tract). The upper trachea contains the larynx, where the vocal cords are found. After passing the larynx, the air moves down the trachea and into the bronchi, which carry air in and out of the lungs. The bronchi are reinforced with cartilage to prevent their collapse and are lined with cilia and mucus-producing cells. Bronchi branch into smaller and smaller tubes known as bronchioles. Bronchioles terminate in little sacs known as alveoli. Alveoli are surrounded by a network of thin-walled capillaries.

When you inhale, muscles in the chest wall contract, lifting the ribs and pulling them, outward, while simultaneously the diaphragm moves downward, both of which contribute to increasing the volume of the chest cavity. This reduces air pressure in the lungs so that air can enter the lungs (inhalation). Exhaling reverses these steps (the diaphragm moves up, the intercostal muscles relax, causing a smaller chest volume, and hence increased pressure, which results in air leaving the lungs). All of these muscles are stimulated by nerves that carry messages from the respiratory control center in the medulla of the brain. Once the air is in the alveoli, gas exchange can occur. At the tissues this involves diffusion of gases from a higher concentration to a lower concentration: oxygen concentration in cells is low (when leaving the lungs, blood is 97% saturated with oxygen), so oxygen diffuses from the blood to the cells when it reaches the capillaries. Carbon dioxide formed in metabolically active cells is much greater than that in the capillaries, so carbon dioxide diffuses from the cells into the capillaries. Carbon dioxide cannot (and certainly should not) be dissolved in the blood, as it is in carbonated drinks. So, water in the blood combines with the carbon dioxide to form bicarbonate ions and hydrogen ions. This removes the carbon dioxide from the blood because the body "packages" the carbon dioxide for transport and eventual passage out of the body. In the alveolar capillaries, bicarbonate combines with a hydrogen ion (proton) to form carbonic acid, which breaks down into carbon dioxide and water. The carbon dioxide then diffuses into the alveoli and out of the body with the next exhalation.

Key Points

The organs of the respiratory system function so that oxygen enters the body and carbon dioxide leaves the body. Gas exchange assists the cells of the body in carrying out cellular metabolism (which uses O_2 and produces CO_2) to make ATP.

- The nose, pharynx, and larynx make up the upper respiratory tract.
- The nose warms and filters incoming air.
- The pharynx transmits air from the nose to the larynx (and is part of the digestive system).
- The epiglottis covers the glottis when swallowing so that only air goes to the larynx (and food enters the esophagus).
- The trachea, bronchi, bronchioles, and alveoli make up the lower respiratory tract.
- The trachea contains cilia to sweep impurities trapped in mucus up to the throat, and it sends air to the bronchi.
- Bronchi, which divide into smaller bronchioles, transmit air into the lungs.
- The lungs lie within the thoracic cavity, and are enclosed by two pleural membranes (that produce fluid to keep the lungs "attached" to the ribcage).
- Alveoli are small sacs surrounded by capillaries that enable gas exchange between inhaled air and the blood. Surfactant is a chemical produced within the alveoli to prevent their collapse (very important in determining if a premature baby will require artificial respiration upon delivery).

- Ventilation is the process of breathing, of getting air into and out of the lungs
- Inspiration moves air into the lungs. The diaphragm and intercostal muscles contract to increase intrathoracic volume. This creates a partial vacuum, allowing air to move into the lungs.
- Expiration moves air out of the lungs. Relaxation of the diaphragm and intercostal muscles decreases intrathoracic volume, and the increased pressure pushes air out of the lungs.
- Tidal volume is the amount of air that moves in and out with each breath.
- Inspiratory reserve is the maximum amount of air that can be inhaled.
- Expiratory reserve is the maximum amount of air that can be exhaled.
- Vital capacity = tidal volume + inspiratory reserve + expiratory reserve.
- Some air can never be expelled from the lungs. This is called residual volume.
- Breathing rate is controlled by the respiratory control center in the medulla oblongata.
- Chemoreceptors (found in the carotid arteries and aortic arch) monitor things like the pH of blood. The more acidic the blood means that more carbon dioxide is being produced. To compensate for this, the rate of breathing increases to get rid of the extra CO_2 and get the breathing rate back to normal—homeostasis!
- Gas exchange depends on simple diffusion (high pressure to low pressure).

Figure Credit

Fig. 10.1: Source: https://en.wikipedia.org/wiki/File:Respiratory_system_complete_en.svg.

Lesson 11

Sustaining Life Functions—Digestion and Nutrition

Objectives

- **List** and state the function of each organ of the gastrointestinal tract.
- **List** the accessory organs and name a function for each.
- **Describe** the structure of the stomach and explain its role in digestion.
- **Describe** the structure of the small intestine and explain its role in digestion.
- **Explain** how carbohydrates, lipids, and proteins are processed by the small intestine.
- **Explain** the functions of the salivary glands, pancreas, liver, and gallbladder during digestion.
- **List** the secretions of the salivary glands, pancreas, liver, and gallbladder.
- **Describe** the structure and function of the large intestine.
- **Discuss** how the body uses nutrients that have been digested and absorbed to power the body for anabolic and catabolic activities.
- **Explain** how energy balance is required for weight maintenance.
- **Describe** how our bodies use energy through metabolism.
- **Explain** the following three types of metabolism: basal metabolic rate, physical activity, and thermic effect of food.
- **Explain** how our bodies regulate hunger.
- **Describe** the differences between hunger and appetite.
- **Read** a nutrition facts label and judge the nutritional value of a food.
- **Distinguish** sound nutritional information from unreliable nutritional information.
- **Describe** a healthy diet and food choices.
- **Explain** why such choices will help prevent health problems.

You have developed specialized structures for ingesting and digesting your food. The digestive system uses mechanical and chemical methods to break food down into nutrients that can be absorbed into the blood. This system encompasses the different physiological systems of digestion, absorption, and nondigestible waste excretion. It includes the alimentary (digestive/GI) tract, accessory digestive organs, and nutrient use.

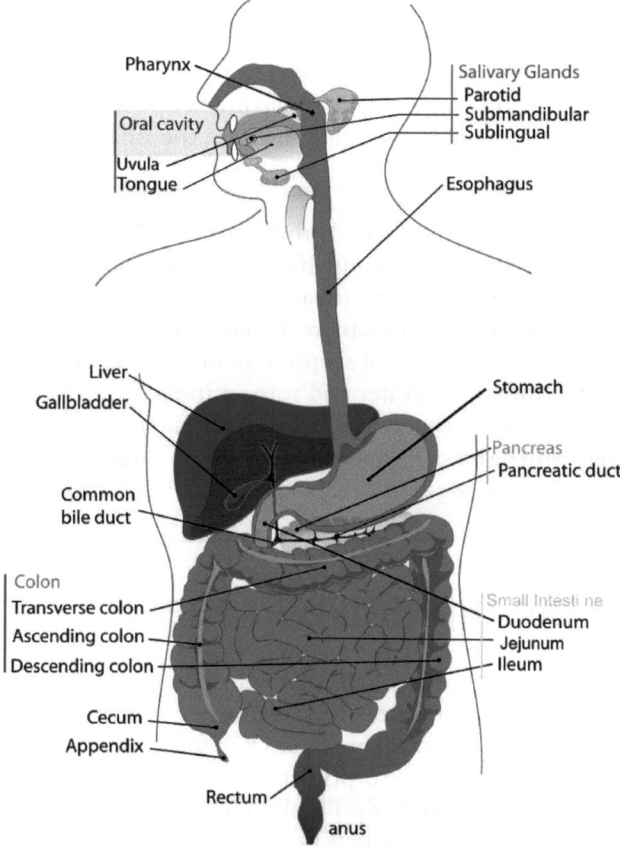

FIGURE 11.1. The Digestive (Gastrointestinal) System

Figure 11.1 exhibits the organs of the digestive system.

Overview

Compartmentalized breakdown of foods

- Enzymes
- Stomach acid

How do we get nutrients from the outside to the inside? How are smaller units absorbed and transported to the cells of your body?

Digestion involves the mechanical and chemical breakdown of food into smaller components that can be absorbed into the body.

The process of digestion involves many organs:

- **Oral cavity:** Secretion of saliva (which contains the carbohydrate digestive enzyme amylase) helps produce a soft, moist bolus of food that can pass down the pharynx and esophagus.
- **Esophagus:** Passageway from oral cavity to stomach
- **Stomach:** Gastric juice and enzymes help break down proteins in food.
- **Small intestine:** Most digestion takes place in the small intestine, where nutrients are absorbed.
  - **Gallbladder:** Where bile (a fluid produced by the liver) is stored before release into the small intestine to emulsify (make water-soluble) fats
  - **Pancreas:** Both an endocrine (insulin and glucagon hormone secreting) and a digestive organ. Secretes pancreatic enzymes and bicarbonate (raises pH) to help with digestion and absorption of nutrients in the small intestine.
  - **Large intestine/colon:** Water and some minerals are absorbed back into the blood. The colon is where most of the bacteria in the GI tract live.
  - **Rectum and anus:** Waste products of digestion are stored and eventually defecated.

Stages in the Digestive Process

Peristaltic movement (by smooth muscles) propels food through the digestive system. Digestive enzymes and other substances are secreted in response to a food stimulus. This enables the mechanical and chemical breakdown of food into small enough molecules to cross the plasma membrane for absorption of these molecules into the body and then be passed along to the rest of the body. Finally, elimination or excretion (removal) of undigested food and solid waste products occurs.

Components of the Digestive System

This system is made of a coiled, muscular tube that goes from the mouth to the anus: mouth, pharynx, esophagus, stomach, small intestine (duodenum, jejunum, and ileum), large intestine (cecum with attached appendix, ascending colon, transverse colon, descending colon, sigmoid colon, rectum), and anus. Accessory organs include the salivary glands, the exocrine pancreas, and the liver and gallbladder.

The Mouth, Pharynx, and Esophagus—Some Mechanical and Chemical Digestion

The mouth provides mechanical breakdown by chewing and chemical breakdown of carbohydrates by amylase from the salivary glands. The mixed food and saliva proceeds down into the pharynx and esophagus, whose smooth muscular contractions (peristalsis) propel food to the stomach. Swallowing moves food from the mouth through the pharynx and into the esophagus (after swallowing lifts the larynx to place the epiglottis over the glottis, preventing food from entering the respiratory tract) on the way to the stomach. This bolus of food is prevented from returning back up from the stomach into the esophagus by a sphincter. Heartburn (gastroesophageal reflux) results from "burning" of the esophagus by the gastric acid that leaks back through the sphincter.

The Stomach—More Mechanical and Minor Digestion

The stomach can hold about one to two liters of chewed food. Cells lining the surface of the stomach secrete gastric hydrochloric acid (HCl), pepsinogen (which is converted to the protein-digesting enzyme pepsin), and mucus (protective coating). These secretions are affected by smells and thoughts of food, as well as endocrine signals. The stomach produces HCl (to lower the pH in the stomach so that pepsinogen can be activated to pepsin) and pepsin (an enzyme that causes the breakdown of proteins into smaller peptides). The stomach also mechanically churns the food until it is liquefied into what is called chyme. The chyme leaves the stomach and enters the duodenum of the small intestine. Alcohol and aspirin are the only ingested substances that may be absorbed from the stomach into the blood. There are also cells that produce mucus to protect the cells from the stomach's HCl. Peptic ulcers may result when the protective mucus fails to protect the inner lining of the stomach. Most peptic ulcers are caused by the bacterium Helicobacter pylori (H. pylori). Other factors affecting development of ulcers include stress and aspirin.

The Small Intestine

The small intestine is where digestion and absorption will be finished. The small intestine is a long, coiled tube (increased surface area). Digestion of proteins and carbohydrates must be completed after being started in the stomach and mouth, respectively. Fats must begin and complete digestion here. The upper part of the small intestine, the duodenum, is where most digestion occurs. The villi, finger-like projections of the small intestine, provide the increased surface area, and also produce enzymes to complete digestion. Absorption also occurs in the small intestine. After food is digested into the small monomers, they can pass out of the small intestine into the circulatory system. Sugars and amino acids go directly into the blood via capillaries in each villus. Glycerol and fatty acids go into the lymphatic system via the lacteal in the villus.

Gastrin, secretin, and cholecystokinin are hormones that regulate digestion. Protein in the stomach stimulates secretion of gastrin, which causes increased stomach acid secretion and movement of the food to the duodenum. Food passing into the duodenum causes the production of secretin, which then stimulates release of secretions from the pancreas. Cholecystokinin (CCK) is released from intestinal cells in response to the presence of fats. This causes the release of bile from the gallbladder and the lipid-digesting enzyme lipase from the pancreas.

The small intestine is aided in digestion and absorption of nutrients by two accessory organs: the pancreas and the liver. Secretions from the pancreas and liver (via the gallbladder) aid digestion in the duodenum. The pancreas secretes digestive enzymes and bicarbonate (to neutralize stomach acid). The liver produces bile, which is stored in the gallbladder and then secreted into the duodenum to emulsify lipids. In addition to aiding digestion, the liver functions with other systems to detoxify the blood, synthesize blood proteins, destroy old erythrocytes and convert hemoglobin into bilirubin, store glucose as glycogen, and produce urea from amino acids being used for fuel (an amino acid minus its nitrogen is a carbohydrate).

In the small intestine carbohydrates are broken down into disaccharides, proteins are digested by proteases (enzymes secreted from the pancreas) into peptides and amino acids, and after bile emulsifies fats, they are broken down into smaller fatty particles by the fat digesting enzyme lipase. Bile is an emulsifier of fats, which allows the ingested fats to mix with the pancreatic lipase and be digested. This fat digestion occurs completely in the small intestine, unlike the digestion of carbohydrates and proteins. Most absorption occurs in both the duodenum and jejunum (second portion of the small intestine) through the intestinal villi. The disaccharides maltose, sucrose, and lactose are broken down by disaccharidase enzymes into monosaccharides that can enter the capillary. Lactose intolerance results from the genetic lack of the enzyme lactase, with resulting gas, bloating, and massive diarrhea. Peptides are digested to amino acids, which can enter the capillary. Gluten sensitivity is the inability to absorb gluten, a protein found in wheat, which then leads to an immunologic inflammatory response. And, because digested fats are water insoluble (hydrophobic), bile salts secreted from the gallbladder help to emulsify fats so that they can pass into the small lymphatic vessels (lacteals) found in the villus.

The Large Intestine

The large intestine, the final portion of the digestive tract, is made up the cecum, appendix, colon (ascending, transverse, descending, and sigmoid), and rectum. Material in the large intestine is mostly indigestible remnants of ingested materials and liquid. Secretions in the large intestine are alkaline mucus to protect tissues and neutralize acids from intestinal bacteria. Water, salts, and vitamins are absorbed here, and the remaining contents of the large intestine form the feces (mostly cellulose, and bacteria, which produce vitamins, such as vitamin K, and bilirubin).

Regulation of Appetite

The hypothalamus in the brain has two centers controlling hunger. One is the appetite center, the other the satiety center. Ghrelin from the stomach stimulates hypothalamic cells to produce neuropeptide Y (NPY) and agouti-related peptide (AgRP). These are both strong orexigenics (substances that stimulate a sense of hunger and desire to eat). Many other substances from the GI tract, and leptin from adipose tissue, stimulate hypothalamic cells to produce proopiomelanocortin (POMC) and cocaine-amphetamine related transcript (CART). These are both anorexigenics (substances that inhibit hunger and the desire to eat), to signal satiation and decrease the desire to eat.

Nutrition

Nutrition deals with the composition of food, its energy content, and nutrients ingested, digested, and absorbed to allow the body to produce energy in the form of

ATP. Macronutrients are foods required in large quantities each day: carbohydrates, lipids, and proteins. Water is an essential nutrient. Despite many varied recommendations for percentages of these macronutrients in the diet, it is most important that they be quality macronutrients, not just a quantity.

Proteins are eaten as a source of amino acids. Proteins are found in meat, milk, poultry, fish, cereal grains, and beans. They are needed for cellular growth and repair. There are 20 naturally occurring amino acids, but you can only synthesize 11 of these. The remaining 9 are considered essential amino acids and must be supplied in the diet. Normally proteins are not used for energy, but during periods of prolonged starvation, muscle proteins can be broken down for energy. Excess protein can be used for energy or converted to fats for storage.

Fats contain the greatest amount of energy. These are found in oils, meats, butter, and plants (such as avocado and peanuts). Two fatty acids, linolenic and linoleic acid, are essential and must be included in the diet (fatty fish, like salmon, is a good source). Dietary fats are also needed to assist in the uptake of the fat-soluble vitamins A, D, E, and K.

Vitamins and minerals are considered micronutrients. Vitamins are organic molecules required for metabolic reactions. You cannot make them, so your body needs to obtain them in trace amounts from the diet. Some vitamins are fat soluble, some are water soluble. Minerals are elements required for normal metabolism, as components of cells and tissues, and for nerve and muscle function. They are obtained from the diet. Iron (for hemoglobin), iodine (for thyroxin), calcium (for bones), and sodium (for nerve message transmission) are some examples of needed minerals.

There is a relationship between nutrients and health. Excess and insufficient intake can cause disease. Nutrition may be a factor in cardiovascular disease, hypertension, and cancer, among others.

Nutrition Labels

- What is a nutrition facts label?
- What are the major components of the nutrition facts label?
  - Serving size and number of servings is based on the amount per one serving, but many items contain multiple servings.
    - Make sure to consider the number of servings you are consuming when looking at the fat, calories, nutrients, and so on.
  - Calories: "Energy in"
  - Fats: Broken down into saturated fat, unsaturated fat, and trans fat
    - Trans fat does not have a percent daily value (%DV) because the goal is to consume as little as possible.
  - Sodium: Too much sodium can lead to high blood pressure.
  - Carbohydrates: Broken down into dietary fiber and sugars. Fiber may keep you fuller longer, and may help your digestive system work properly.
    - Try to limit foods with added sugar.
  - Protein: Most Americans get plenty of protein.
    Vitamins/minerals: Amounts of various micronutrients included in the product.

- Percent daily value: The percentages listed on the far right represent how much of the total amount the FDA recommends of each component we should eat each day per serving.

- Ingredients: Ingredients are listed in order from the greatest to least amount included in the product. It is a good place, for example, to make sure the food doesn't contain trans fats (listed as "partially hydrogenated oil") and does contain whole grains (whole grain wheat, oats, etc.).

Health Claims

- FAT FREE: The product has fewer than 0.5 grams of fat per serving.
- LOW FAT: The product has 3 grams or fewer of fat per serving.
- REDUCED or LESS FAT: Product has at least 25 percent less fat per serving than the full-fat version.
- LITE or LIGHT: The product has fewer calories or half the fat of the nonlight version.
- CALORIE FREE: The product has fewer than 5 calories per serving.
- LOW CALORIE: The product has 40 calories or fewer per serving.
- REDUCED or FEWER CALORIES: The product has at least 25 percent fewer calories per serving than the nonreduced version.
- Front label deception: Terms like "fortified," "enriched," "added," "extra," and "plus" usually mean the food has been altered or processed in some way.
- "Fruit drinks" usually means little or no real fruit and a lot of sugar.
- "Made with wheat," or "rye," or "multi-grains" implies that it's a good source of whole grains, but unfortunately, doesn't tell you how much whole grain is actually in the product. Look for the word "whole" before the grain to ensure that you are getting a 100 percent whole-grain product.
- "Natural" or "made from natural" means the manufacturer started with a natural source. Once processed, the food may not resemble anything "natural."
- "Sugar-free," "sugarless," or "no added sugar" tells you nothing about sugar derivatives or sugar substitutes, which may yield just as many calories as table sugar and may be more harmful than sugar.

Key Points

Eating involves the ingestion of food, digestion (mechanical and chemical breakdown of ingested food) of dietary carbohydrates, fats, and proteins to small monomeric subunits, and finally absorption of these small nutrient subunits. Elimination of nondigested foods removes waste products from the digestive tract.

- One layer of the wall of the GI tract is muscular (smooth muscle for peristalsis).
- The mouth (where food is ingested and mechanical and chemical digestion of carbohydrates, by amylase, starts), is followed by the pharynx (back of the throat,

- which can contain air or food), and then the esophagus (tube that takes food to the stomach) to begin the GI tract.
- Swallowing must block food from entering the respiratory tract and allow food to only enter the esophagus.
- Peristalsis is the rhythmic contraction that pushes food along the digestive tract.
- Sphincters (muscles that surround tubes in the digestive and urinary tracts) are found between many components of this pathway to prevent materials from proceeding before they are ready. (Regurgitation of stomach contents back into the esophagus causes heartburn.)
- The stomach stores food while it churns to mechanically digest foodstuff, and also mixes food with acidic secretions (HCl and pepsin, for protein digestion to begin).
- Very little absorption occurs in the stomach.
- The small intestine contains and receives enzymes to complete digestion.
- The duodenum gets bile from the gallbladder, while the pancreas provides multiple enzymes.
- Lipase digests fat after being emulsified by bile.
- Pancreatic amylase completes digestion of carbohydrates.
- Pancreatic trypsin and other protein enzymes complete protein digestion.
- The small intestine absorbs sugars (monosaccharides), amino acids, glycerol, and fatty acids.
- Villi allow nutrients to be absorbed into the blood, though fats are first absorbed into lymphatic vessels, via the lacteal, before being processed by the liver into particles that can exist in the aqueous blood.
- The accessory organs of digestion (pancreas, liver, and gallbladder) provide secretions to aid digestion in the duodenum.
- Secretions of digestive substances are controlled by both the nervous system (parasympathetic nervous system stimulates gastric secretion when you see/smell food—"rest and digest") and by digestive hormones—stomach (gastrin) and duodenum (CCK and secretin).
- The large intestine (cecum, colon [ascending, transverse, descending, and sigmoid], and rectum) ends at the anus.
- The large intestine absorbs some water, salts, and vitamin K (from bacteria) and B_{12} (if bound by intrinsic factor), and forms and stores the feces until they are defecated.
- Nutrients are components of food necessary for physiological functions to occur properly.
- Carbohydrates are used to meet energy needs. Complex are better than refined carbohydrates, which have a high glycemic index.
- Complete proteins are foods with all of the essential amino acids (animal products) and are necessary. Strict vegetarians can accomplish this by eating combinations of legumes, grains, vegetables, seeds, and nuts to get all of them.
- Fats and oils (saturated, unsaturated, and trans fatty acids), and cholesterol are dietary lipids.
- Only linoleic and linolenic acids are essential dietary fats.
- Current thinking is that lipids should be used sparingly because they may contribute to plaque on blood vessel walls. But, is this a cause or an association?
- Cholesterol can be transported in the blood by High Density Lipoprotein (HDL a scavenger) and within Low Density Lipoprotein (LDL, a transporter) particles (transport to cells for normal use—cell membranes, steroids, etc.).

- Vitamins and minerals are necessary for proper enzyme function.
- Vitamins are organic compounds that the body cannot produce on its own, so need to be part of a healthy diet.
- Vitamins may be fat or water soluble.
- Vitamins A, C, and E are considered antioxidants (protect cell from damage by free radicals).
- Vitamin D (one of the fat-soluble vitamins—A, D, E, K) is necessary for proper calcium absorption, and bone maintenance.
- Calcium helps maintain strong bones.
- Sodium intake may be associated with hypertension in some people.
- A "good" dietary recommendation is to "Just Eat Real Food," in moderation.

> Free radicals: The body is under constant attack from oxidative stress. Oxygen in the body splits into single atoms with unpaired electrons. Electrons like to be in pairs, so these atoms, called free radicals, scavenge the body to seek out other electrons so they can become a pair. This causes damage to cells, proteins and DNA. Free radicals are associated with human disease, including cancer, atherosclerosis, Alzheimer disease, Parkinson's disease, and many others. Antioxidants, like vitamins A, C, and E, help to keep free radicals in check and preserve the life cells under such attack.

Figure Credit

Fig. 11.1: Source: https://pixabay.com/vectors/digestive-system-human-digestion-41529/.

Waste Management Functions

LESSON 12

Objectives

- **Summarize** the functions of the urinary system.
- **Identify** the organs of the urinary system and state their functions.
- **Identify** the structures of a nephron and state the function of each.
- **Summarize** the three processes involved in the formation of urine.
- **Summarize** how the kidney maintains the water–salt balance of the body.
- **State** the purpose of ADH and aldosterone in water–salt homeostasis.
- **Explain** how the kidneys assist in the maintenance of the pH levels of the blood.

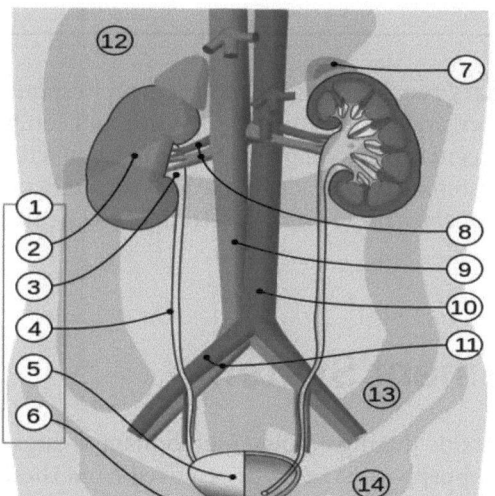

1. Urinary system
2. Kidney
3. Renal pelvis
4. Ureter
5. Urinary bladder
6. Urethra
7. Adrenal gland
8. Renal artery and vein
9. Inferior vena cava
10. Abdominal aorta
11. Common iliac artery and vein

Shaded:
12. Liver
13. Large intestine
14. Pelvis

FIGURE 12.1. The Urinary System

Figure 12.1 is an overview of the structures (BOXED 1, 2, 4, 5, and 6) that makeup the urinary system. Within each kidney are millions of nephrons (see Figure 12.2)—the functional units of the kidney, where filtration, reabsorption, and secretion occur to rid the body of wastes and produce urine.

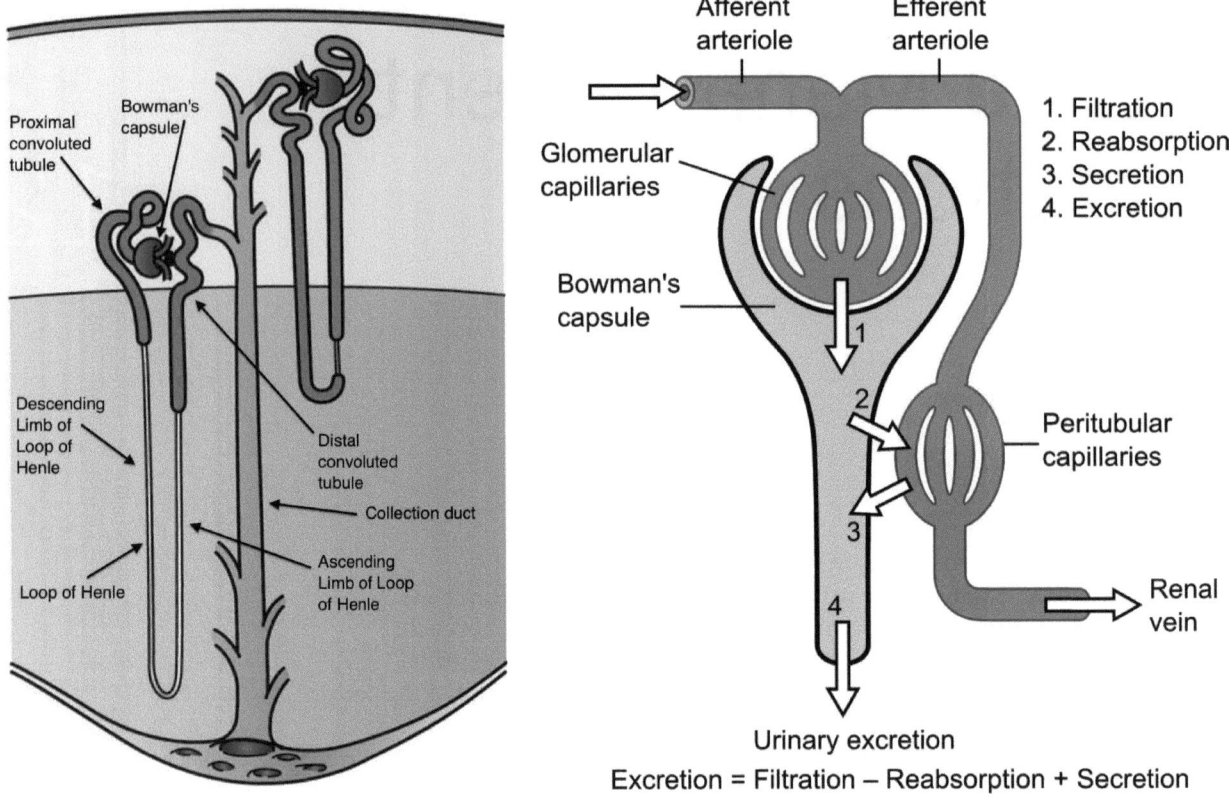

FIGURE 12.2. The Nephron

The kidneys are part of the urinary/renal system. The kidneys are responsible for filtering the blood, eliminating wastes, and maintaining the volume and composition of the blood. They also play a role in regulating blood pressure and pH. You will explore the structure and function of the excretory system, deamination of amino acids, and the three basic processes (filtration, reabsorption, and secretion) of the nephron (the functional unit of the kidney).

The Waste Management System

Cells produce water and carbon dioxide during aerobic metabolism of sugars, fats, and proteins. Nitrogen, sulfur, and phosphorous must be removed from these larger molecules to prepare them for use as energy sources. The large amounts of cellular wastes diffuse out of cells and into the blood. You have a specialized system to remove these waste products from cells, and then transport them in the blood to the kidneys, where they will eventually be delivered for removal from the body. These excretory systems not only remove metabolic waste products, but also regulate the chemical composition of body fluids by retaining the amounts of water, salts, and nutrients needed for homeostasis.

Nitrogenous wastes are a byproduct of protein metabolism. Amino groups (NH_2) are removed from amino acids, and then combine with a hydrogen ion (proton) to form ammonia (NH_3), so that the remainder of the amino acid molecule can be used for energy production (ATP). Ammonia is very toxic and usually must be excreted. Ammonia is converted to urea by the liver, and then transferred into the blood and excreted by the kidneys.

Water and salt balance are managed by the excretory system. You have internal concentrations of salt that must be regulated to stay within certain concentration limits. Osmotic regulation involves preventing water loss from the body and preventing salts from diffusing into the body from the environment (remember the discussion of hypo-, iso-, and hyper-tonic solutions). You have various means to reduce water loss, also.

The Human Excretory System

The human excretory, or urinary system is made up of the kidneys, ureters, bladder, and urethra. The kidney's functional unit is the nephron. The blood is filtered and eventually excreted as urine by each kidney. Urine leaves the kidneys via the ureters, collects in the bladder, and leaves through the urethra. The nephron consists of a glomerular capsule (Bowman's capsule) full of capillaries, and the start of a long tubule. Blood flows into the nephron through an afferent (going to) arteriole, branches into the glomerular capillaries, and then exits through the efferent (going from) arteriole. The efferent arteriole then wraps around the nephron's tubules, where materials not filtered can still be exchanged with the nephron, if necessary, to produce the final urinary product. High capillary pressure due to a larger diameter in the afferent arteriole compared to the efferent arteriole causes water and solutes from the blood to **filter** into the capsule. This filtrate then goes through the proximal convoluted tubule (PCT), down and up the loop of Henle, and then into the distal convoluted tubule (DCT) before emptying into a collecting duct. Beneficial fluids and solutes (nutrients like glucose and amino acids, ions, etc.) are **reabsorbed** into the surrounding peritubular capillaries that follow the nephron's tubule, whereas toxins and other substances (nitrogenous waste products, etc.) to be removed either stay in the filtrate, or are **secreted** from blood into the filtrate. Thus, the nephron has three functions: glomerular **filtration** of water and solutes from the blood; tubular **reabsorption** of water and important, necessary molecule transfer back into the blood; and tubular **secretion** of ions and other waste products from the surrounding capillaries, which are transferred back into the filtrate at the distal tubule.

Kidney function in homeostasis:

- Excrete toxic metabolic byproducts (urea, ammonia, and uric acid)
- Maintain extracellular ionic balance
- Maintain extracellular fluid volume
- Maintain extracellular pH and osmolarity

Hormones and the Kidneys

Water reabsorption is affected by antidiuretic hormone (ADH, also called vasopressin) in a negative feedback fashion. ADH is released from the posterior pituitary gland in the brain after the hypothalamus sends it signals to release the hormone into the blood. ADH increases "pores" in the collecting ducts, allowing for more water absorption from the kidneys back into the blood. This movement of water back into the blood concentrates the urine. If too much fluid is present in the blood (hypervolemia; possibly from the effect of alcohol or caffeine), sensors cause a decrease in the amount of ADH released into the blood. This increases the amount of water excreted by the kidneys, producing large quantities of more dilute urine (diuresis).

Aldosterone, a hormone secreted by the adrenal cortex, affects the exchange of sodium from the filtrate with potassium in the blood. When sodium levels in the blood decrease, aldosterone is released into the blood, causing more sodium to be reabsorbed from the filtrate into the blood. This happens to allow water to follow the sodium back into the blood. Renin (part of the renin-angiotensin-aldosterone system) is released by the kidneys (juxtaglomerular apparatus) to control aldosterone production and release.

Anemia is another condition detected by the kidneys. They have special cells that will secrete erythropoietin in response to decreased oxygen carrying capacity (anemia!). This hormone travels to the red bone marrow of the long bones, sternum, and iliac crest to stimulate red blood cell production (erythropoiesis), which will alleviate the anemia. Erythropoietin is also used by some athletes to illegally enhance their performance (blood doping).

Kidney Stones and Other Renal Problems

At times, excess waste products (calcium, uric acid, etc.) may crystallize out to form kidney stones (nephrolithiasis). They may grow and become painful. They may require surgery or ultrasound treatments. Some stones are small enough to be forced into the urethra and passed from the body, a very painful situation. Other kidney problems may be brought about by infection, environmental toxins, and genetic diseases. Some of these serious kidney problems may need to be treated by dialysis. Dialysis is a process whereby a machine acts as a kidney. Kidney transplants may be necessary if dialysis shows a need for a more permanent solution to poor kidney function.

Key Points

The urinary/renal system contributes to maintaining salt, water, and pH homeostasis. This system processes the blood prior to excretion of wastes and performs other important functions that contribute to homeostasis.

- The kidneys are found in the back, just below the bottom of the ribs, where they produce urine to rid the body of metabolic waste products.
- Urea is a waste product of amino acid metabolism and is the main nitrogenous waste product of metabolism (others are creatinine, from creatine-P, and uric acid, from metabolism of nucleotides).
- The ureters transport urine from the kidneys to the bladder, where urine is stored until it leaves the body through the urethra.
- The nephron is the functional unit of the kidney.
- Each nephron receives blood from an afferent arteriole at the glomerulus.

- High osmotic pressure filters the blood from the glomerular capillaries to the glomerular (Bowman's) capsule.
- Filtration allows small molecules (water, wastes, and nutrients) to move from the glomerulus to the inside of the glomerular capsule.
- An efferent arteriole leaves the capsule and surrounds the remaining nephron.
- From the glomerular capsule, the filtrate enters the proximal convoluted tubule (PCT), where reabsorption of filtered substances mainly occurs.
- Tubular reabsorption at the PCT removes the nutrients and water from the proximal convoluted tubule and sends them back into the blood of the peritubular capillaries.
- The filtrate then travels through the loop of Henle (descending and ascending) before entering the distal convoluted tubule (DCT). Secretion occurs along these two structures.
- Tubular secretion allows certain substances, like hydrogen ions, creatinine, and drugs that were not adequately filtered, to be sent from the blood back into the filtrate at the distal convoluted tubule.
- Most of the water found in the filtrate is reabsorbed into the blood before urine leaves the body, depending on the posterior pituitary hormone, antidiuretic hormone (ADH).
- Finally, the filtrate leaves the nephron via the collecting duct (where water balance occurs in response to the hormone ADH).
- The kidneys regulate the water and salt balance (sodium and potassium) to maintain normal blood volume and blood pressure.
- The kidneys release renin (activating the renin-angiotensin-aldosterone system) in response to low blood pressure. This leads to aldosterone secretion by the adrenal cortex and also leads to vasoconstriction caused by angiotensin II.
- Aldosterone from the adrenal cortex causes an exchange of blood potassium ions (K^+) with filtrate sodium ions (Na^+) in response to the juxtaglomerular apparatus secretion of renin when blood pressure is low.
Antidiuretic hormone (ADH) acts when osmolarity increases (too much "stuff" in the blood, which needs to be diluted).
- The normal blood pH should be maintained between 7.35 and 7.45.
- The kidneys help the body manage acids and bases to keep blood pH close to 7.4.
- A buffer is a substance/mechanism that helps to resist a change in pH.
- An important buffer system in the blood is the carbonic acid/bicarbonate buffer system.
- The respiratory system contributes to this buffer system by changing the respiratory rate to alter the carbon dioxide in the blood.
- Erythropoietin (EPO) is released by the kidneys to increase red blood cell production.
- The kidneys also synthesize vitamin D_3.
- In diabetes mellitus, glucose appears in the urine if the blood glucose level exceeds a threshold.
- Wherever glucose goes, water goes, thus causing increased urination and resulting dehydration (which stimulates increased drinking of fluids because of the thirst center in the hypothalamus).

Summary

Urine production involves filtration in the glomerulus and Bowman's capsule, reabsorption in the proximal convoluted tubule, and tubular secretion mainly in the loop of Henle and distal convoluted tubule.

Components and functions of the nephron:

- Bowman's capsule/glomerulus: Filters blood
- Proximal convoluted tubule: Reabsorbs most of the water, salts, glucose, and amino acids
- Loop of Henle: Maintains the concentration gradients
- Distal convoluted tubule: Salt balance and tubular secretion of H^+ ions, potassium, and certain drugs
- Collecting ducts: Manage water balance

Figure Credits

Fig. 12.1: Copyright © Jordi March i Nogué (CC BY-SA 3.0) at https://commons.wikimedia.org/wiki/File:Urinary_system.svg.

Fig. 12.2a: Copyright © by Holly Fischer (CC BY 3.0) at https://commons.wikimedia.org/wiki/File:Kidney_Nephron.png.

Fig. 12.2b: Copyright © by Madhero88 (CC BY 3.0) at https://commons.wikimedia.org/wiki/File:Physiology_of_Nephron.png.

Male and Female Sexual Functions

LESSON 13

Objectives

- **List** the functions of the reproductive system in humans.
- **Identify** the structures of the male reproductive system and provide a function for each.
- **Describe** spermatogenesis.
- **Summarize** how hormones regulate the male reproductive system.
- **Identify** the structures of the female reproductive system and list a function for each.
- **List** the stages of the ovarian cycle and explain what is occurring in each stage.
- **Describe** the process of oogenesis.
- **Summarize** how estrogen and progesterone influence the ovarian cycle.

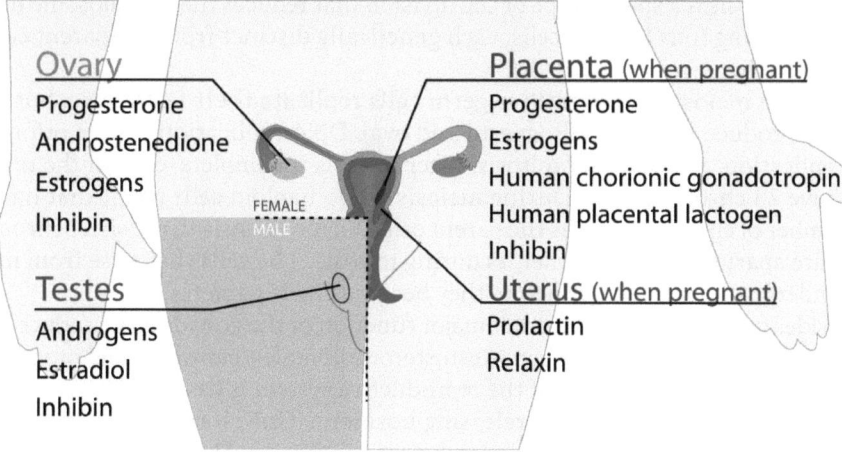

FIGURE 13.1. Male and Female Reproductive Systems

Figure 13.1 shows the male and female reproductive systems' sex organs, which work together towards sexual reproduction and reproductive secondary sex characteristics. Unlike most organ systems in the body, the "opposite" sexes often have significant differences. These differences allow for a combination of genetic material between two individuals.

You will learn about the reproductive systems of males and females and how they are specialized to produce gametes (sperm and eggs). This system also enhances the chances of fertilization and implantation for "survival of the species." Fertilization will usually be followed by healthy development of the embryo and fetus into a new human being. You will explore the structure and function of the reproductive system, spermatogenesis and oogenesis, and menstrual and ovarian cycles.

Reproductive System and Gametogenesis

Interestingly, the reproductive systems of both sexes share the same basic organization: The gonads produce gametes (sperm or eggs) whose nuclei fuse during fertilization. They also produce steroid hormones essential for reproductive secondary sexual characteristics. The genitalia, the external reproductive structures, include the penis and scrotum in males, and the vulva in females. There are various ducts that connect the gonads to the genitalia, and accessory glands to enhance the probability of successful fertilization. The ducts allow for the passage of sperm from the male into the female ducts, where fertilization may occur. These ducts are also where the gametes prepare for fertilization, and, in females, they also create an environment suitable for fertilization to occur. The uterus allows the resulting fertilized egg (zygote) to develop into an embryo, and then a fetus, until birth of the offspring.

The creation of gametes is called spermatogenesis in males, and oogenesis in females. The gametes begin as diploid cells, containing 23 pairs of chromosomes (46 chromosomes). Through mitosis, they give rise to more germ cells with that same number of chromosomes. Eventually, the germ cells go through the two phases of meiosis to form gametes. Meiosis is a special type of cell division that reduces the chromosome number by half, creating four haploid cells, each genetically distinct from the parent cell that gave rise to them.

Thus, in meiosis, the primitive germ cells replicate their DNA once, but divide twice to produce haploid cells (sperm and ova). DNA replication occurs prior to the first replication, as occurs for mitosis. After meiosis is complete, each of the resulting cells have 23 chromosomes. During meiosis I, two haploid cells result that have half the number of chromosomes as the parent cell. During meiosis II, the sister chromatids segregate apart from one another, as during mitosis. The cells that arise from meiosis may undergo special changes before they become viable gametes.

Besides gametogenesis, the other major function of the gonads is to produce steroid sex hormones, including androgens (testosterone in males, estrogens and progesterone in females). Hormonal control of the reproductive system is basically the same in both men and women: gonadotropin-releasing hormone (GnRH) from the hypothalamus causes the release of follicle-stimulating hormone (FSH) and luteinizing hormone (LH) from the anterior pituitary. These two hormones then travel to the gonads, where they have their sex-specific effects.

Endocrine Control

The hypothalamus produces bursts of gonadotropin-releasing hormone (GnRH). GnRH travels to the anterior pituitary. GnRH causes the release of two peptide hormones from the anterior pituitary: follicle-stimulating hormone (FSH) and luteinizing hormone (LH). FSH acts on the gamete-producing cells to regulate gametogenesis (spermatogenesis in the testes of males, and oogenesis in the ovaries of females). LH acts on the endocrine/hormone-producing cells to stimulate release of steroid sex hormones (testosterone in males, estrogens and progesterone in females) as well as ovulation during the menstrual cycle in females.

Feedback control over production of steroid sex hormones is controlled by the gonadal hormones secreted into the circulation, eventually returning to the hypothalamus and pituitary of the brain to suppress GnRH, FSH, and LH production.

Anatomy of the Male Reproductive System

- External genitalia (penis and scrotum)
- Urethra (common passage for semen and urine)
- Testes (male gonads produce sperm, which travel from testes to urethra via the vas deferens)
- Accessory glands (prostate gland, seminal vesicles, and bulbourethral glands, which add secretions to sperm to produce semen)

Sperm production takes place within the seminiferous tubules of the testes, where the germ cells (spermatogonia) develop into sperm by the process of meiosis (I & II). This development is regulated by FSH from the anterior pituitary via stimulation of Sertoli cells to start the process of spermatogenesis. The other function of the testes is to produce testosterone (done by Leydig cells). These cells are under the influence of luteinizing hormone (LH) from the anterior pituitary, resulting in the production of testosterone. Testosterone negatively feeds back to the anterior pituitary to decrease release of LH.

Anatomy of the Female Reproductive System

- External genitalia (labia majora, labia minora, and clitoris—together termed the vulva)
- Vagina connects vulva to cervix of the uterus (which serves to receive sperm during sexual intercourse and carry the fetus until birth)
- Cervix (neck of uterus that protrudes into upper end of the vagina)
- Walls of the uterus contain thick, smooth muscle called the myometrium. It serves to expel the fetus during labor. The inner lining of the uterus (endometrium) proliferates in preparation for implantation of the zygote and is then shed during each menstrual cycle.

- Two Fallopian tubes (where fertilization typically occurs) to connect the uterus to the ovaries (female gonads).

Egg Production Takes Place within the Ovaries

In the male, all stages of sperm production take place throughout life. In the female, each stage takes place during specific periods. Starting with about seven million germ cells, or oogonia, in the embryonic ovary, only about a half million proceed through meiosis I to become primary oocytes before birth. And these are ALL of the ova a female will have for life.

After puberty, primary oocytes begin to develop (a few each month). Each will be contained in a follicle and be under the influence of FSH (similar to its the influence on spermatogenesis in males).

One or more ovarian follicles will rupture at the middle of the menstrual cycle, releasing an egg into the fallopian tube. Hormone production (estrogen and progesterone) in the ovaries occurs within the follicles. Production of these two hormones varies over the course of the menstrual cycle:

- The first day of menstruation is usually labeled as the beginning of the cycle. Increasing levels of FSH stimulate development of several follicles in the ovaries. These early follicles produce estrogen. One of these follicles will continue to develop.
- Estrogen levels continue to rise as the follicle develops. Progesterone then begins to rise. Estrogen causes negative feedback on both GnRH and gonadotropin (FSH/LH) secretion.
- Around day 14 of the cycle, the anterior pituitary changes its response to high estrogen levels. Estrogen now has a positive feedback effect.
- The switch to positive feedback causes an LH surge. High LH levels initiate ovulation, causing the follicle to rupture and the egg to be released.
- After ovulation, the follicle becomes the corpus luteum (yellow body), which produces progesterone. The corpus luteum only remains viable for about 14 days, unless fertilization occurs. If there is no fertilization of the ovum by a sperm, both estrogen and progesterone levels taper off, causing the endometrium to be shed. This also relieves the negative feedback on GnRH and FSH production, and the cycle begins again. If fertilization does occur, the corpus luteum continues producing progesterone to maintain the integrity of the uterus until it can take over production of progesterone from the corpus luteum. How does the corpus luteum know to continue producing progesterone? (How does it know pregnancy has occurred?) Human chorionic gonadotropin (hCG) is produced by the uterus!

Key Points

The reproductive system is both similar and different between males and females. In both, the reproductive organs (testes in males, ovaries in females) produce gametes (sperm and eggs). They also both produce sex hormones (testosterone in males, estrogens and progesterone in females).

- The testes, like the ovaries, begin development inside the abdominal cavity until about embryonic week seven.
- If testosterone is released and sensed by receptors, the testes descend into the scrotum (this helps regulate the temperature of the testes).
- If there is no testosterone stimulus, the individual develops female reproductive structures.
- Mitosis is like cloning—producing two exact copies of one original cell. It occurs during growth and repair of tissues. Meiosis produces inexact copies of the original cell. This allows for variety/differences in offspring. Meiosis occurs in the testes of males and in the ovaries of females. During meiosis in humans, the number of chromosomes is halved; from 46 chromosomes (diploid) to 23 chromosomes (haploid).
- In males, testes produce sperm that travel via the vas deferens to the urethra. Secretions are added from the seminal vesicles, prostate gland, and bulbourethral glands.
- Spermatogenesis takes place in the seminiferous tubules of the testes.
- Mature sperm have a head, a middle piece, and a tail (flagellum).
- The head has a cap called the acrosome (secretes enzymes to penetrate the egg).
- Testosterone is secreted by cells that lie between the seminiferous tubules (interstitial/Leydig cells).
- Gonadotropin-releasing hormone (GnRH) is secreted by the hypothalamus. It then travels to the anterior pituitary to release follicle-stimulating hormone (FSH; spermatogenesis) and luteinizing hormone (LH; testosterone).
- Testosterone exerts negative feedback on the hypothalamus and anterior pituitary.
- Testosterone causes male secondary sex characteristics (greater muscular development, more red blood cells, deeper voice, increased length of long bones).
- The female gonads are the ovaries, which produce female sex hormones (estrogen, progesterone) and eggs (ova).
- Estrogen is mostly responsible for secondary sex characteristics.
- An egg released by an ovary is swept towards, and enters, a Fallopian tube/oviduct (thanks to the fimbriae), which leads to the uterus, where implantation (into the endometrium, or uterine lining) may occur.
- From the uterus, the cervix is encountered next, which then enters the vagina.
- The endometrial lining is shed each menstrual period if the released egg has not been fertilized.
- Hormone levels follow a distinct monthly cycle in females.
- The ovarian cycle drives the uterine cycle.
- The ovarian cycle begins with development of the follicle (FSH from the anterior pituitary stimulates development of an oocyte-containing follicle, and secretion of estrogen and progesterone), followed by ovulation (day 14, LH surge), and development of the corpus luteum (secretes mostly progesterone, but also some estrogen).
- If pregnancy does not occur, the corpus luteum "dies," progesterone levels fall, and the endometrium is shed, leading to the start of a new cycle (which begins with menstruation).
- Menstruation (days 1–5) starts the menstrual (uterine) cycle and is due to low levels of estrogen and progesterone.
- During days 6–13, the endometrium thickens under the influence of estrogen.
- Ovulation occurs with a surge of LH, mid-cycle (day 14).

- The secretory phase (days 15–28) allows the endometrium to continue to increase in thickness and vascularity (under the influence of progesterone).
- If pregnancy does not occur, the uterine lining (endometrium) is shed and the cycle begins again.
- If fertilization does occur, the zygote/embryo implants in the endometrium of the uterus.
- A placenta develops from fetal and maternal tissues to nourish the embryo.
- It produces human chorionic gonadotropin (hCG) to prevent degeneration of the corpus luteum, allowing the uterine lining to be maintained.

Figure Credit

Fig. 13.1: Source: https://en.wikipedia.org/wiki/File:Endocrine_reproductive_system_en.svg.

Protective Immune Functions

LESSON 14

Objectives

- **List** examples of the body's innate defenses.
- **Summarize** the events in the inflammatory response.
- **Explain** the role of an antigen in the adaptive defenses.
- **Summarize** the process of humoral immunity and list the cells involved in the process.
- **Summarize** the process of cell-mediated immunity and list the cells involved.
- **Distinguish** between active and passive immunity.
- **Explain** what causes an allergic reaction.

This lesson will introduce you to the wonderful and fascinating world of defense by the human immune system against invading pathogens.

The Lymphatic and Immune Systems

The lymphatic system consists of lymphatic vessels, lymph nodes, and lymphoid organs. The functions of this system include absorption of excess fluid (with its eventual return to the blood), absorption of fat (in the villi of the small intestine), and immune system functions.

The primary lymphoid organs are the bone marrow and thymus, whereas the tonsils, lymph nodes, and spleen are among the many secondary lymphoid organs. Bone marrow contains tissue that produces lymphocytes. B lymphocytes (B cells) mature in the bone marrow. T lymphocytes (T cells) mature in the thymus gland. The thymus secretes the hormones thymosin and thymopoietin, which cause pre–T cells to mature (in the thymus) into T cells. Other white blood cells, such as monocytes and granulocytes (neutrophils, basophils, and eosinophils) are also produced in the bone marrow and function in the process of phagocytosis, allergies, and parasitic infections. Lymph nodes contain many lymphocytes and macrophages residing along the lymphatic vessels. The spleen is a lymph node-like organ, which is larger than a lymph node and filled with blood. The spleen is a reservoir for blood, and filters blood and lymph fluid.

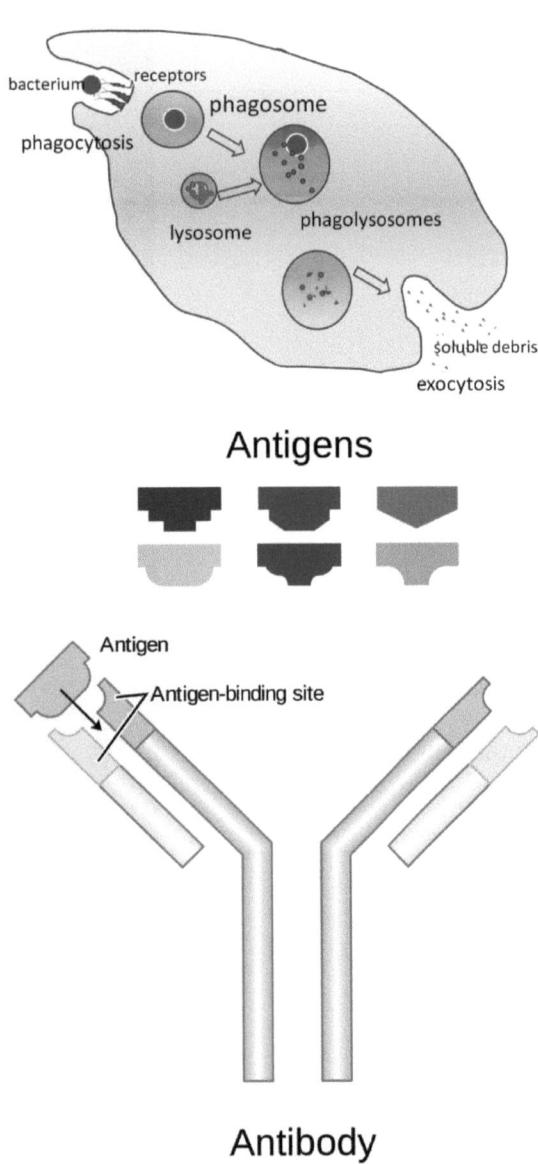

FIGURE 14.1. Phagocytosis and Antibodies

Figure 14.1 shows the nonspecific immune protective process of phagocytosis and the specific defense molecules called antibodies.

Immunity

Immunity is the body's capability to repel foreign invading substances and cells (recognized as antigens). Nonspecific responses are the first two lines of defense. Specific responses are the third line of defense. Nonspecific responses (enzymes, phagocytosis, inflammatory response) block the entry and spread of pathogens (disease-causing antigens). Antibody-mediated and cell-mediated responses are the specific responses. The immune system defends against disease-causing agents, transplant rejection, blood transfusion reactions, and diseases resulting from "hyperimmune" responses to "normal" antigens (autoimmune diseases and allergies). Lack of normal third-line defense from human immunodeficiency virus (HIV) infection results in suppressed immune function, which may progress to acquired immunodeficiency syndrome (AIDS).

General Defenses

First barriers to entry by antigens are the skin and mucous membranes. As long as there are no breaks, the skin is a large, passive barrier to infectious agents such as bacteria and viruses. Any antigens that enter the eye encounters tears, which secrete destructive enzymes, and pathogens that enter through the mouth encounter saliva, which also secretes enzymes that breakdown bacterial cell walls. Skin secretes chemicals to slow the growth of bacteria. Membranes lining the respiratory, digestive, urinary, and reproductive tracts secrete mucus to slow the invasion until the pathogens can be dealt with.

If microorganisms penetrate skin or the lining of the respiratory, digestive, or urinary tracts, the inflammatory response occurs. Damaged cells release chemicals that increase capillary blood flow into the affected area (causing heat and redness). The heat is unfavorable for microbial growth, promotes healing, and, along with the chemicals, promotes movement of white blood cells to the area (chemotaxis). Capillaries leak fluid into the tissues, causing the infected/injured area to swell (stimulating nociceptors, which signal pain). Finally, monocytes (a type of phagocyte) clean up dead microbes, cells, and debris.

The inflammatory response is usually strong enough to slow, or even stop, the spread of disease-causing agents (viruses, bacteria, and fungi). If this is unsuccessful, the complement system and third line of immune response kick into action.

The complement system proteins (produced in the liver) bind to bacteria, drilling holes in their membrane, causing fluids and ions to move in. This is followed by swelling and bursting of the cell. The complement system enhances the inflammatory response, and works with the second-line immune response by tagging the outer surface of invaders for attack by phagocytes.

Interferon is a chemical produced by virally infected cells. It alerts nearby cells to prepare for a virus, and resist all viral attacks.

Specific Defenses

The immune system is usually able to distinguish and identify cells as self or non-self.

Antibody-mediated (humoral) immunity is managed by B cells and their antibodies. Antibody-mediated reactions defend against invading bacteria. Cell-mediated immunity is managed by T cells. Cell-mediated immunity defends against cells that have been infected by viruses, and also kills cancer cells.

Antigens can be any foreign molecule that causes antibody production. Immune protection may result from the production of antibodies specific to a given antigen (antibody generator). Antibodies bind to the antigens on invaders. Antibody-mediated immunity stages involve antigen detection, activation of helper T cells, and antibody production by B cells.

These specific immune responses attack specific invaders and are more effective than the nonspecific response. This line of defense has a memory component that improves response time, and increases the quantity of defenders, when another attack by the same invader occurs.

Macrophages

Macrophages are large phagocytic white blood cells that look for foreign (non-self) antigens, viruses, or microbes. The macrophages will engulf and destroy them. Small fragments of the antigen are also displayed on the outer surface of the macrophage to communicate with helper T cells regarding which branch of specific immunity to activate (humoral or cell-mediated). Helper T cells can activate B cells to produce plasma cells, which produce antibodies. Cytotoxic (or killer) T cells destroy cells infected with a virus or cancer. Memory T cells, like memory B cells, remain in the body until the next encounter with the antigen. A second exposure to the same antigen produces a more massive and faster response.

B Cells

B cells divide to form plasma cells and memory B cells. Plasma cells make and release antibodies into the blood. These antibodies bind to specific antigens. Antibodies are also known as immunoglobulins. There are five classes of immunoglobulins: IgG, IgA, IgD, IgE, and IgM. Antibodies are Y-shaped molecules composed of two identical long chains (heavy or H) and two identical short chains (light or L). Antibodies recognize and bind to a site on the antigen, leading to its destruction.

T Cells

A cell infected with a virus will display viral antigens on its plasma membrane. Killer T cells recognize the viral antigens and attach to that cell's plasma membrane. The T cells secrete proteins (perforin—perforates!) that punch holes in the infected cell's plasma membrane. The infected cell dies, and is removed by phagocytes. Killer T cells may also bind to cells of transplanted organs, causing rejection.

The Immune System and Memory

Resistance to certain diseases occurs after having had them once, or being immunized against that pathogen. This results from production of memory B and T cells during the first exposure to the antigen. The secondary response is also the basis for vaccination. Vaccination (experimented with by Edward Jenner in the first vaccination) stimulates antibody production and formation of memory cells without causing the disease. Vaccines are made from killed or weakened pathogens to cause antibody production. Active immunity develops after an illness or vaccine. Passive immunity is the type of immunity when the individual is given antibodies to combat a specific disease. Passive immunity doesn't last long, and produces no memory.

Blood Types, Rh, and Antibodies

There are many known antigens on the surface of blood cells. These form the blood groups/types. In a transfusion, the blood groups of the recipient and donor must be matched. If they don't match properly, the recipient's antibodies will attack the incompatible antigens, causing agglutination (clumping) of the transfused cells. This will lead to blockage of circulation through capillaries and will produce serious, or even fatal, results. ABO blood types are determined by a gene. There are three alleles (gene variants), A, B, and O. Proteins produced by the A and B alleles are antigenic. Individuals with blood type A have the A antigen on the surface of their red blood cells, and antibodies to the B antigen in their plasma. People with blood type B have the B antigen on their blood cells and antibodies against A antigens in their plasma. Individuals with type AB blood have both antigens (A and B) on their cell surfaces and no antibodies for either in their plasma. Type O individuals have neither A nor B antigens on their red blood cells but antibodies to both A and B are in their plasma.

The Rh (rhesus monkey) blood group is made up of Rh positive (Rh$^+$) individuals, who make the Rh antigen, and Rh negative (Rh$^-$) individuals, who do not. Hemolytic disease of the newborn (HDNB) results from Rh incompatibility between an Rh$^-$ mother and Rh$^+$ fetus. Rh$^+$ blood from the fetus enters the mother's blood during birth, causing her to produce Rh antibodies. The first child is usually not affected. However, subsequent Rh$^+$

fetuses cause a massive secondary immune response by the mother's immune system. To prevent HDNB, Rh⁻ mothers are given an Rh antibody during pregnancy with an Rh⁺ fetus, and all subsequent Rh⁺ fetuses, in order to prevent a memory response.

Organ Transplants and Antibodies

Success of organ transplants and skin grafts requires a match of antigens on the cells of both donor and recipient. Because of the large number of genes involved, no two non-identical twins will have the same identical antigens. Identical twins have a 100% match.

Allergies and Disorders of the Immune System

The immune system can overreact, causing allergies or autoimmune diseases. A defective immune system can result in disease and death.

Allergies result from immune system hypersensitivity to weak antigens that do not cause an immune response in most people. Allergens, substances that cause allergies, include dust, molds, pollen, cat dander, certain foods, and some medicines (such as penicillin). After exposure to an allergen, some people make IgE antibodies, instead of IgG antibodies. Subsequent exposure to the same allergen causes a massive secondary immune response that releases a lot of IgE antibodies. These bind to mast cells (tissue basophils), which release histamine. This starts the inflammatory response. In some individuals, the histamine release causes life-threatening anaphylaxis or anaphylactic shock.

The immune system usually distinguishes self from non-self. The immune system learns the difference between its own cells and foreign cells. Autoimmune diseases result when the immune system attacks and destroys its own cells and tissues. Type I diabetes, Grave's disease, multiple sclerosis, systemic lupus erythematosus, and rheumatoid arthritis are examples of autoimmune diseases.

Immunodeficiency diseases result from the lack, or failure, of part of the immune system. Affected individuals are susceptible to diseases that normally would not affect most people. Genetic disorders, Hodgkin's disease, cancer chemotherapy, and radiation therapy can cause immunodeficiency diseases. Severe combined immunodeficiency (SCID—"boy in the bubble") results from a complete absence of the cell-mediated and antibody-mediated immune responses. Affected individuals suffer from a series of seemingly minor infections and usually die at an early age. Some SCID patients have an adenosine deaminase (ADA) deficiency. There are clinical trials whereby these individuals undergo gene therapy to provide them with normal copies of the defective gene.

Acquired immunodeficiency syndrome (AIDS) results from the destruction of helper T cells by the human immunodeficiency virus (HIV). As with viruses in general, the viral genome becomes incorporated into the patient's DNA for months or years. Gradually, the number of CD_4 helper T cells, the master communicator for the specific immune response, declines. The immune system may eventually fail. Premature death results from a series of rare diseases (such as fungal pneumonia and Kaposi's sarcoma, a rare cancer) that overwhelm the body and its compromised immune system.

Summary

Lymphatics/Lymph Nodes

- Composed of lymphatic vessels and lymphatic cells
- Screen body tissues for foreign antigens
- Lymph nodes house white blood cells called **macrophages/lymphocytes** that recognize and attack foreign antigens present in lymph.

Antigens

- Foreign molecules trigger a specific immune response.
- Include components of **bacterial cell walls**, as well as proteins of **viruses**, and other pathogens
- Food and dust can also contain antigenic particles.
- Enter the body by various methods:

    - Through breaks in skin and mucous membranes
    - Direct injection, as with a bite or needle
    - Through organ transplants and skin grafts

First Line of Defense—Nonspecific

Innate immunity is **nonspecific**, meaning that these lines of defense (first and second) work against a wide range of pathogens.

- Structures (skin), chemicals (stomach acid, lysozymes), and processes (phagocytosis) that work to *prevent pathogens from entering the body.*
- Includes the **mucous membranes** of the respiratory, digestive, urinary, and reproductive systems

Second Line of Defense—Nonspecific

- Operates when pathogens get past the first line of defense by penetrating skin or mucous membranes
- Cells (phagocytes), antimicrobial chemicals (lysozymes), and processes (inflammatory response), but no physical barriers
- Many of these components are contained in, or originate in, the blood.

Components of the Second Line of Defense

Leukocytes—Phagocytosis

Phagocytes ingest and destroy foreign matter, such as microorganisms or debris.

- Extracellular killing by leukocytes

  - Nonspecific chemical defenses
  - Inflammation
  - Fever

Nonspecific Chemical Defenses

- Lysozyme and cytokines (interferons and interleukins)
- Enhance phagocytosis

  - Complement—opsonization (increase "palatability," like ketchup for your french fries!)
  - Some attack pathogens directly
  - Some enhance features of nonspecific resistance

Inflammation

- Nonspecific response to tissue damage
- Damaged cells release histamines, which increase vasodilation.
- Heat, swelling, pain
- **Fever**

Third Line of Defense—Acquired/Specific

- The body's ability to recognize and defend itself against distinct invaders

  - "Memory" allows it to respond rapidly to additional encounters with a pathogen.

- Two types of specific immunity:

  - **Naturally acquired:** Immune response against antigens encountered in daily life
  - **Artificially acquired:** Response to antigens introduced via vaccine

Antibodies

- Also called immunoglobulins (Ig); 5 classes: IgG, IgM, IgA, IgD, IgE
- Proteinaceous molecules that bind **antigens**
- Considered part of the **humoral immune response** (Body fluids like lymph and blood were once called humors.)

How Antibodies Work

- Some act as **opsonins**—identify antigens for phagocytes, then stimulate phagocytosis.

- Some work as **antitoxins**—they neutralize toxins, like those that cause diphtheria and tetanus.
- Some cause **agglutination** (clumping together) of bacteria, making them less likely to spread.

Lymphocytes—Where Antibodies Are Produced

Two main types:

- **B cells** mature in bone marrow, then congregate in lymph nodes and spleen.
- **T cells** mature in thymus.

T Lymphocytes (T Cells)

- Produced in red bone marrow and mature in thymus
- Circulate in the lymph and blood in order to migrate to the lymph nodes
- Part of the **cellular immune response** (aka cell-mediated immune response)
- These cells act directly against various antigens:
    - Endogenous invaders (intracellular pathogens inside the body's cells)
    - Abnormal body cells, such as cancer cells
    - Types:
        - **cytotoxic** or **killer** T cells (T_C)
            - Destroy compromised body cells
        - **helper** T cells (T_H)
            - Activate T and B cells

B Lymphocytes (B Cells)

Activated B lymphocytes produce either

- **plasma cells,** which make antibodies against a pathogen; or
- **memory cells,** which remember a given pathogen for faster antibody production in future infections.

Key Points

The lymphatic system is made up of lymphatic vessels and primary and secondary lymphoid organs. Its functions include absorbing excess interstitial fluid; absorbing fats via intestinal lacteals; and supplying lymphocytes for immune system function—defending the body against pathogens (foreign, invading substances/antigens).

- The lymphatic organs include primary organs (red bone marrow and thymus) and secondary organs (tonsils, appendix, lymph nodes, and spleen).

- Red bone marrow produces B cells (lymphocytes), which mature in the bone marrow.
- T cells (lymphocytes) are also produced in the bone marrow, but mature in the thymus.
- The thymus produces the thymic hormones thymosin and thymopoietin.
- Secondary lymphoid organs, like the spleen, contain many lymphocytes.
- The spleen is also involved in filtering the blood.
- Immunity involves innate (inborn) and adaptive defenses.
- Innate defenses (nonspecific) protect against any pathogen.
- Adaptive defenses (specific) are effective against specific infectious agents.
- Nonspecific defense includes both physical (skin, cilia) and chemical (sweat, enzymes, low pH) defenses to block infection.
- The inflammatory response is characterized by redness, heat, swelling, and pain.
- Histamine from damaged tissues and mast cells causes the redness and swelling.
- Increased temperature assists phagocytosis by neutrophils and macrophages.
- Swelling stimulates the sensation of pain.
- Complement proteins assist innate immunity by attracting phagocytes (chemotaxis) and enhancing phagocytosis (opsonization).
- The conclusion of the action of complement proteins is to form a membrane attack complex (MAC), which drills holes in membranes of bacteria, causing them to burst.
- Interferons are produced by virus-infected cells to help noninfected cells prepare for possible viral attack.
- When pathogens get by innate defenses, the adaptive defense kicks in.
- Adaptive defenses respond to antigens (molecules recognized by the immune system as non-self).
- Adaptive defense depends on B and/or T lymphocytes.
- Activated B cells become both plasma cells (to produce antibodies and memory B cells to remember previously encountered antigens).
- There are five different classes of antibodies, depending on a particular part of the antibody molecule: IgG, IgM, IgA, IgD, and IgE.
- Every plasma cell derived from a given B cell secretes the same antibodies against its specific antigen (monoclonal antibodies).
- T cells are cytotoxic, and attack virus-infected and/or cancer cells.
- For a T cell to recognize an antigen, the antigen must be presented by an antigen-presenting cell (APC; often a macrophage).
- A T cell, called a helper T cell, is the intermediate between the APC and either B cells (for humoral immunity) or T cells (for cell-mediated immunity).
- Helper T cells produce cytokines—chemicals that enhance immune cell responses.
- HIV, the virus that causes acquired immunodeficiency syndrome (AIDS), affects helper T cells.
- This makes HIV-infected individuals more susceptible to opportunistic infections, which would not typically harm most people.
- Memory T cells, like memory B cells, remember, and are ready to react to a previously encountered pathogen.
- Immunity can be brought about naturally or artificially.
- Active immunity develops when you are infected with a pathogen.
- Vaccines can also cause active immunity, without having the disease.
- Either type of active immunity is long lived because of memory B and T cells.

- Passive immunity involves giving an individual antibodies produced somewhere else to fight a disease (nursing passes antibodies from mother to child, gamma-globulin shots are injections of pre-formed antibodies).
- Passive immunity is temporary, and produces no memory.
- Sometimes the immune system hyper-responds.
- Allergic responses occur when the immune system reacts to substances not normally antigenic (allergens).
- Instead of forming IgG antibodies to the allergen, IgE is produced.
- IgE binds to mast cells (tissue basophils), which contain and release histamine (antihistamines are effective treatment for many allergic symptoms).
- Transplant rejection involves cytotoxic T cells destroying foreign, non-self tissue (the transplant).
- Autoimmune diseases occur when the immune system fails to recognize self-antigens as self, leading to attack of the body's own cells (rheumatoid arthritis, systemic lupus erythematosus, myasthenia gravis, multiple sclerosis).

Figure Credits

Fig. 14.1a: Copyright © by Graham Colm (CC BY-SA 3.0) at https://commons.wikimedia.org/wiki/File:Phagocytosis2.png.

Fig. 14.1b: Source: https://commons.wikimedia.org/wiki/File:Antibody.svg.

Concluding Remarks

LESSON 15

Our bodies are fascinating. Human biology is amazingly complex. It can be hard to grasp at first when you consider how many complicated processes are carried out by our bodies every second. Whether you are a bio expert or know absolutely nothing about the human body, it's useful to know human biology basics in order to make informed decisions that will help keep your body working. Plus, knowledge of the human body can be quite interesting, as exemplified by the following facts.

- The human body has 12 systems.

They are the integumentary system, the skeletal system, the muscular system, the nervous system, the endocrine system, the cardiovascular system, the respiratory system, the digestive system, the lymphatic and immune systems, the urinary system, and the reproductive system.

All these systems work together to ensure that our bodies work correctly. The integumentary system protects the body from outside damage. The skeletal system gives our bodies a framework and supports the other systems. The muscular system allows us to move. The nervous system transmits signals through the body and controls voluntary and involuntary actions. The endocrine system produces hormones that regulate metabolism, growth and development, tissue function, sexual reproduction, sleep, and mood. The cardiovascular (or circulatory) system transports blood, oxygen, and nutrients throughout the body. The respiratory system enables us to take in oxygen and expel carbon dioxide as we breathe. The digestive system takes in and processes food. The urinary system expels waste. The reproductive system allows us to have sex and children. The lymphatic system connects the lymph nodes in our bodies and helps the circulatory and immune systems. The immune system fights infection.

- There are four blood groups: A, B, AB, and O.

Your lettered blood type is determined by which antigens are found on your red blood cells. Antigens are foreign substances that activate an immune response and control what enters and exits a cell. Antibodies are blood proteins produced in response to encountering an antigen. Each blood group can be either positive or negative. The +/- part of a person's blood type is determined by the presence (or absence) the antigen called the Rh (Rhesus) factor.

- Our DNA is stored in 23 pairs of chromosomes within the nucleus of every cell of the body.

Each cell, other than sperm and ova, has a full set of chromosomes, which contain all the genetic material needed to determine the makeup of our entire bodies. That's why cloning of animals can be done with just one cell. All the genetic material that defines us is inside each cell of our body.

- Our immune system fights off infection mostly using antibodies and macrophages.

Antibodies fight infection by killing the virus or foreign bacteria, while macrophages are white blood cells that surround and contain the foreign cells (or other objects) to prevent the spread of disease.

- There are more nonhuman cells in our body than human ones.

There are ten times more bacteria cells in our bodies than our own human cells. These bacteria are harmless or even help us perform key bodily functions, such as digestion.

- We have more than five senses—each has its own sensory organ or special receptors.

We typically think of the traditional five senses of touch, taste, hearing, vision, and smell, but our bodies can also sense many other things:

1. Balance
2. Temperature
3. Proprioception (spatial body awareness—touch your nose with your eyes shut)
4. Pain

- Nearsightedness and farsightedness are caused by defects in the shape of our eyeballs.

Nearsightedness, or myopia, is caused by a greater curve in the cornea of the eye or by an elongation of the eyeball. Farsightedness, or hyperopia, is caused by a corneal curve that is too small or by a having a short eyeball. Some evidence indicates that nearsightedness is genetic.

- The red color of our blood is caused by iron, within hemoglobin, binding oxygen.

Many people may think that blood is red simply because of all the iron in it. The red color is created because the iron is bound in a ring of atoms in hemoglobin called porphyrin. This structure has a shape that makes the blood appear red. When oxygen is bound to the porphyrin ring, it changes the shape, making our red blood cells appear as an even more vivid shade of red.

- The brain works harder while we are asleep than during the day when we are awake.

Many people may think that sleep helps the brain rest, but our brains are busier during sleep than while we are awake. When we sleep and dream, our brains carry out important functions that they cannot perform while focusing on movement and conscious thought. During sleep, our brains process things we learned and emotions we felt during waking hours, and create memories.

- The liver has over five hundred functions.

Our liver doesn't just filter toxins from the blood. It does much more to keep our bodies healthy. Some of its other important functions include creating bile to help in the emulsification of fat, producing cholesterol, regulating blood clotting, processing hemoglobin, and so much more.

Printed by Libri Plureos GmbH in Hamburg, Germany